Mathematical Analysis of Environmental System

Jun Tanimoto

Mathematical Analysis
of Environmental System

 Springer

Jun Tanimoto
Kyushu University
Hakata, Japan

Tanimoto Kyouju no (Doryoku Sureba) Darenidemo Wakaru Kankyo System no
Suuri Kaiseki Kiso, © 2012 Jun TANIMOTO, All rights reserved
Original Japanese edition published in 2012 by Kyushu University Press

ISBN 978-4-431-54621-4   ISBN 978-4-431-54622-1 (eBook)
DOI 10.1007/978-4-431-54622-1
Springer Tokyo Heidelberg New York Dordrecht London

Library of Congress Control Number: 2013955041

Printed on acid-free paper

Springer is part of Springer Science+Business Media (www.springer.com)

Foreword

I am very pleased and proud to announce the launch of the Green Asia Lecture Book Series. Green Asia (Global Strategy for Green Asia) is one of the Programs for Leading Graduate Schools promoted by the Ministry of Education, Culture, Sports & Technology in Japan, in which we aim to establish a science and engineering leadership training program that promotes environmental and energy innovation to reach out from Asia to the rest of the world. One of the outstanding features of Green Asia as an advanced educational program is that all lectures are offered in English. Hence, the publication of cutting-edge textbooks is one of the most important, visible and tangible outputs of the Green Asia Program. Each of the volumes in the lineup deals with essential theories, fundamentals, practical applications or up-coming topics, all of which are actually used in the program lectures. It will be wonderful if our publication project can bring all of the brilliant content and approaches produced in the Green Asia Program to a worldwide audience.

Fukuoka, Japan

Professor Akira Harata, Dr. Eng.
Director of Green Asia Education Center &
Head Coordinator of Advanced Graduate Program
in Global Strategy for Green Asia, Kyushu University
Professor, Interdisciplinary Graduate School of
Engineering Sciences, Kyushu University

Preface

For more than 15 years, I have been lecturing about environmental engineering to master's degree students specializing in engineering and science. Probably because the school I work for is an interdisciplinary graduate school, the background the students acquired during their undergraduate years ranges widely, from mechanical, civil, and architectural engineering to pure science such as physics and applied mathematics, and even to architectural design. On the question of what a graduate student who starts studying environmental issues must learn, I would like to say it is not so widely diverse but, rather, is narrow. It comprises fundamental knowledge of how one mathematically builds the so-called physical balance equations on transferring heat, mass, and momentum, which are usually dealt with in heat and mass transfer theory and fluid dynamics, and also how one achieves practical solutions through a series of numerical procedures, which may be crucially important when he or she becomes a working engineer. I have noticed that, unfortunately, there is not a very appropriate book for students at the master's degree level, while many specialized books for research fellows including Ph.D. students have been published. This is because the fundamentals span several fields, as mentioned above. Therefore, this book brings together all the topics I believe that students at the master's level should know to start studying environmental subjects.

The course for which I am responsible was revised a couple years ago. The new lecture series, entitled "Mathematical Analysis of Environmental System," is given entirely in English. Ironically, I have found it necessary to prepare a Japanese book to make it possible for Japanese students to understand and to be able to learn on their own even if they can understand none of what I am saying in the classroom. Now, to my great pleasure, it has been possible to publish an English version. That is exactly how this book came about. Readers of the whole text will, I hope, appreciate the usefulness of this book.

I sincerely thank my colleagues at the Interdisciplinary Graduate School of Engineering Sciences (IGSES), Kyushu University, for their support.

Fukuoka, Japan   Jun Tanimoto

Contents

Chapter 1
Environmental Systems and Analysis Methods

Abstract In this chapter, we discuss the definition of an environmental system referred in this book and explain the structure of each chapter.

Keywords Human–environmental–social system

1.1 Environmental Systems

A system exists as a collection of elements, all of which are connected organically to form an aggregate of elements that collectively demonstrate an overall function; we assign the word for "system" to describe this. If an environmental system is interpreted literally, since any system involved with the environment can be called so, there is a lot of variety in it.

This variety arises from interactions between different environments (e.g., natural, human, and social environments) and differences in spatial scale (i.e., from the microscopic world weaved by microorganisms to the global environment as a whole; Fig. 1.1). To get to the crux of an environmental problem, we must observe and consider diverse phenomena as an integrated environmental system, taking into consideration all interactions between the different systems and scales (Fig. 1.1). Accordingly, we have coined the phrase *human–environmental–social system* to encompass these diverse phenomena.

As an example, let us turn our attention to the problem of urban heat islands. The presence of these heat islands can be attributed to the effects of the drastic land-use modification[1] that has occurred due to urbanization and the increase in the density of energy consumption. The latter refers primarily to the energy dissipated through

---

[1] This is a combination of the effect of the increased roughness caused by the construction of buildings and the influence of the impaired functions of water surfaces and green spaces (i.e., decrease the evaporative capacity), which typically return solar radiation to the atmosphere through the conversion of sensible heat to latent heat.

J. Tanimoto, *Mathematical Analysis of Environmental System*,
DOI 10.1007/978-4-431-54622-1_1, © Author 2014

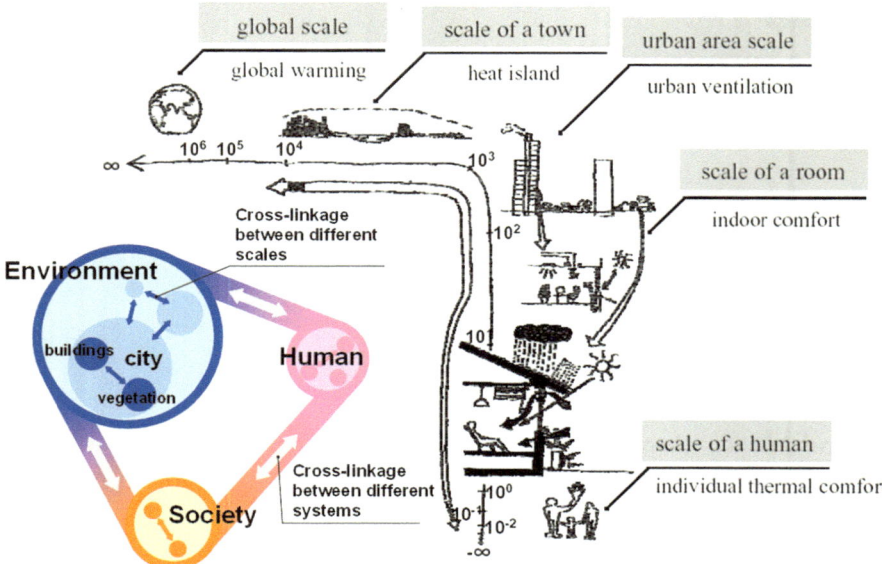

Fig. 1.1 Wide range of spatial scales over which environmental systems act, and the concept of human–environmental–social system

human activity, which generally occurs in towns and cities, such as the use of cars and the heating and cooling of buildings. Such processes are ultimately governed by the second law of thermodynamics; therefore, all energy consumed is eventually expelled into urban air as exhaust heat, increasing the temperature in towns and cities. For example, let us consider summer, the hottest time of the year. In combination with the heat island effect, the high temperatures prompt residents to turn on their air conditioning, which means that the thermal gain (totaling with the electrical power that is input to the air conditioner) is expelled to the environment from the external unit; this causes the air temperature in towns and cities to rise further, resulting in a need for even further cooling. Such a situation worsens the efficacy of cooling and increases energy consumption. If we wish to study this process holistically, just an examination of the thermofluid phenomena in the urban atmosphere would be insufficient. Thus, to accurately reproduce the cycle of cause and effect described above, a concept is required that bridges the gaps between the various systems and scales referred to in Fig. 1.1. Moreover, we must consider a number of factors, such as the types of buildings within the urban area and the type of cooling employed on a room-to-room level, in the analysis.

In accordance, attacking environmental problems (which also represent important social issues) head on is essential. However, to achieve this, we must first achieve the installation of extensive equipment and the establishment of suitable analysis methods. It is unreasonable to think that humans, not being God, could possibly freely take control of this situation as we would wish. With this in mind, it is hoped that environmental studies will continue to advance in the future.

Now, let us define what constitutes an environmental system according to the topics discussed in this book. We can define an environmental system as an organic assembly of physical elements that are expressed by parameters that exhibit spatiotemporal variation. Specifically, we are concerned with determining the spatiotemporal trajectories of the primary physical variables of environmental systems in order to predict the behavior of these systems and design appropriate measures to manage them. That is, we wish to elucidate the dynamics of physical systems and their temporal evolution. To investigate these systems quantitatively, we can evaluate them a priori in situations where appropriate mathematical models have already been developed[2]; otherwise, numerical solutions can be obtained. Typically, mathematical models describing the dynamics of physical systems are expressed as differential equations that include temporally and spatially variable parameters. The meaning of a priori, above and in subsequent instances, is that the differential equation is solved analytically. However, this is only possible in cases where the available mathematical models are exceptionally simple and where specific initial and boundary conditions are satisfied; thus, adopting an a priori solution is difficult. Where possible, the original mathematical models are digitized and solutions are found numerically. In such a way, this is the point where a numerical solution makes an appearance. It is important to note that surprisingly good results can be achieved in the rare instances for which a deductive approach can be applied. Although the results of numerical solutions derived under specific initial and boundary conditions can be dismissed simply as innocuous numerical digital data,[3] a priori (deductive) solutions offer important insights into the behaviors of systems because they incorporate expressions to describe both causes and effects in environmental systems. Meanwhile, the derivation of numerical solutions requires considerable computing power owing to the volume of digitized data that must be processed. Moreover, to obtain a numerical solution, it is important to develop mathematical notation that can be easily processed by computers. In this respect, vector–matrix operations play a critical role; for this reason, a thorough knowledge of linear algebra is essential for the study of environmental systems.

1.2 Structure of This Book

In this book, we provide a thorough discussion of the various analytical methods used in the investigation of environmental systems. In Chap. 2, we describe some examples of environmental systems, particularly linear systems. Here, we adopt the nonsteady heat conduction equation, which will be familiar to many readers, and advance this theory on the basis of a case in which the temperature field over a

---

[2] This refers to mathematical models that describe the physical phenomena of a system.

[3] More specifically, its shortcomings are that, when changing the given conditions, the results cannot be known by analogy, and a numerical solution must be done. This does not provide a good viewpoint.

semi-infinite ground plane is analyzed. A linear environmental system can be represented by the system equation of state, which typically has a vector–matrix representation. We provide a detailed explanation of the nature of this convenient equation of state for a system, with a good outlook, and specifically the setup of the numerical calculation. Moreover, we discuss the stability of numerical solutions and describe the finite element method, which is often used for space discretization. We also provide a detailed explanation of the finer points of the programming involved in room temperature fluctuations and thermal load calculations.

In Chap. 3, we explain the concepts of multiple regression analysis and the least squares solution in an attempt to highlight the power and ease of use of handling mathematical models with vector–matrix representations in conducting quantitative analysis. Because this section does not have a direct bearing on environmental systems, readers who are a little rushed may wish to omit it. However, the concept of the generalized inverse matrix, which is utilized in many important applications, is also presented here; therefore, we recommend that you take the time to study this section if possible.

Chapter 4 discusses analytical techniques used for nonlinear environmental systems. For the deductive approach, which we do not touch upon in Chap. 2, we explain methodological characteristics and provide an example. In fact, the example presented is actually a game, although not one in the traditional sense of games (e.g., video games, arcade games, mahjong, Japanese playing cards, or pinball, although these are related). Rather, this section refers to the game theory of applied mathematics, which abstracts the decision-making of humans. It is possible to model the environmental problems that we touched upon previously as a social dilemma game in which the environment is considered to be public property. In cases where an egocentric decision-making entity makes choices about public property from which everyone can benefit (e.g., shared grazing land, fisheries, or village forests), nobody will initiate a coordinated effort. Thus, overfishing will leave fisheries depleted, grazing land will be abandoned, and village forests will become bare hills. Regardless of the framework applied, the consideration of whether any concerted action will arise can be investigated using evolutionary game theory. As indicated by the description of this method as "evolutionary," the proposed frameworks encompass the concept of time and can be perceived as examples of nonlinear environmental systems.

The authors have taught the abovementioned content for 15 terms to first-year students of a Master's course in a postgraduate school. As such, no fancy or esoteric content is included, and the material should not be viewed with trepidation or anxiety. Although we would not refer to the material presented as "a piece of cake," or "something to do before breakfast" as we say in Japanese, the content should certainly be manageable after you have had a good lunch!

I imagine that, combined with a moderate amount of self-study, no more than a week of careful reading should be required. In fact, the content is sufficiently basic that the astute reader should be able to read through it in half a day even half asleep! I wish the reader the best of luck with the material provided.

Chapter 2
Linear Systems Analysis Methods

Abstract In this chapter, linear systems analysis is described in detail using a representative example. We consider a semi-infinite soil thermal field whose fundamental equation is an unsteady-state heat transfer equation. First, we discuss the physical implications of the fundamental equation, the concept of discretization, and how continuum space is discretized. Next, system state equations based on vector matrix notation are derived. This book describing time discretization for system state equations will help the readers understand the processes of numerical analysis. In the second half, giving examples of changes in single room temperature and thermal load computation, specific programming methods are described in detail. We also discuss the stability of discretized numerical solutions, and introduce the finite element method (FEM) commonly used for space discretization.

Keywords Control volume method • Discretization • System state equation • Unsteady-state heat transfer equation • von Neumann stability analysis

2.1 Unsteady-State Heat Transfer Equation

The one-dimensional unsteady-state heat transfer equation, describing the temporal and spatial evolution of θ [°C] in terms of time t[s] and position x[m], is as follows:

$$C_p\rho \frac{\partial \theta}{\partial t} = \lambda \frac{\partial^2 \theta}{\partial x^2}, \tag{2.1}$$

where C_p is specific heat [J/(K kg)], ρ is density [kg/m^3], and λ is thermal conductivity [W/(m K)]. The thermal conductivity indicates the ease of heat transfer through the material. $C_p\rho$ [J/(K m^3)] is the volumetric specific heat.

The ratio of thermal conductivity to volumetric specific heat is the thermal diffusivity[1] [m^2/s], which may be expressed as

$$a = \frac{\lambda}{C_p\rho}.$$
(2.2)

Equation (2.1) is analogous to Newton's equation of motion $ma = f(m\ dv/dt = f)$. In the heat equation, the temporal velocity change is replaced with a temperature change and the volumetric specific heat represents *thermal mass*. As mentioned below, the right side of Eq. (2.1) is the net heat flux acting on an element. This quantity is equivalent to the force acting on an object in Newton's law. In other words, the conductive heat induces a temperature change $d\theta$ within time dt in an element with heat mass $C_p\rho$ per unit volume. This is analogous to force f producing a velocity change dv within time dt for a particle of mass m.

Here we restrict our discussion to the one-dimensional problem, but the three-dimensional version of Eq. (2.1) is solved similarly:

$$C_p\rho\frac{\partial\theta}{\partial t} = \lambda\left(\frac{\partial^2\theta}{\partial x^2} + \frac{\partial^2\theta}{\partial y^2} + \frac{\partial^2\theta}{\partial z^2}\right).$$
(2.3)

Using Nabla

$$\nabla = \left(\frac{\partial}{\partial x}, \frac{\partial}{\partial y}, \frac{\partial}{\partial z}\right) = \mathbf{i}\frac{\partial}{\partial x} + \mathbf{j}\frac{\partial}{\partial y} + \mathbf{k}\frac{\partial}{\partial z},$$

and the Laplacian

$$\Delta = \frac{\partial^2}{\partial x_1^2}, \frac{\partial^2}{\partial x_1^2} + \cdots + \frac{\partial^2}{\partial x_n^2},$$

we can obtain the following:

$$C_p\rho\frac{\partial\theta}{\partial t} = \lambda\nabla^2\theta$$
(2.4)

$$C_p\rho\frac{\partial\theta}{\partial t} = \lambda\nabla\theta.$$
(2.5)

In the derivation of Eqs. (2.1) and (2.3), a few empirical facts should be assumed. Let us consider two separated points within a material of thermal conductivity λ and assume that a temperature difference occurs, as shown in Fig. 2.1. Note that the

---

[1] The *diffusivity or diffusion coefficient* has the same unit [m^2/s] as for other transport phenomena other than heat (e.g., the molecular diffusivity coefficient). In any phenomenon, the diffusivity coefficient (diffusivity) essentially represents the transport efficiency, which is the constant of proportionality in terms of the relationship between the transferred flux and the concentration gradient imposed by a potential difference (described later). This universal relationship is referred to as Fick's law.

Fig. 2.1 Fourier's law

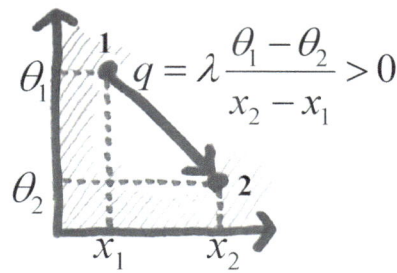

Fig. 2.2 Heat flux flowing
in and out of a micro-
hexahedron

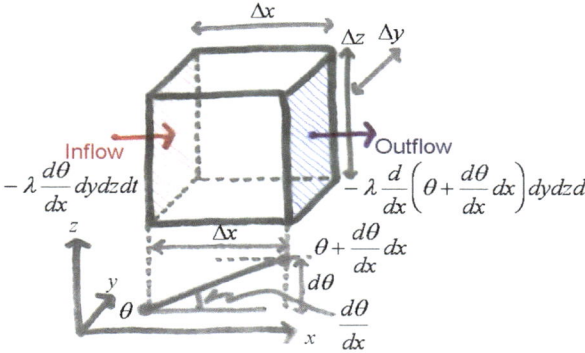

origin of the position coordinate system is to the left. Heat flows from the higher
temperature point 1 to the lower temperature point 2. Intuitively, it may be understood that the larger the temperature difference between the points, the greater the
heat flow. To the end, the heat flux is proportional to the first power of the temperature difference. It is considered that the heat flux reaching point 2 decreases with
increasing separation from point 1. Mathematically, this fact can be expressed that
heat flux is inversely proportional to distance. Finally, it may be considered that the
heat flux depends on the type of material. In fact, a large amount of heat flows through
materials such as metals, but less heat flows through insulation materials. Thus, the
constant of proportionality is referred to as the heat conductivity λ. As inferred from
the above descriptions, the amount of heat flowing by conduction, or the conductive
heat flux q [W/m^2], is expressed by the equation shown in Fig. 2.1. Fourier's law is
derived assuming that distance and temperature difference are infinitesimal.

$$q = -\lambda \frac{d\theta}{dx}. \tag{2.6}$$

It should be noted that the prefix negative sign of the thermal conductivity can be
attributed to the fact that the subscript in the equation of Fig. 2.1 is reversed in the
denominator and the numerator.

Now consider a three-dimensional micro-hexahedron buried within the material,
as shown in Fig. 2.2.

First, the heat flux flowing from the left interface to the micro-hexahedron within time dt is estimated. This means that only the component of heat flow in the x direction is included. Because the heat flux is expressed by Fourier's equation, we just need to multiply the area and time to find the value of heat inflow into the left face of the box shown in the figure. Similarly, the heat flux flowing out of the right interface is estimated. The question we ask is as follows: Given that the temperature at x is θ, what is the temperature at $x + \Delta x$? Take a look at the temperature gradient in the x direction drawn directly under the micro-hexahedron at the bottom of Fig. 2.2. Because the temperature gradient is approximately linear, the temperature at the outflow interface at an infinitesimal distance Δx can be written as $\theta + \frac{d\theta}{dx} \cdot x$ (where $\Delta x = dx$). This expression is exactly $y = $ (intercept) + (gradient) $\cdot x$, which you might be familiar since primary school days. Substituting this expression into the term of temperature of a Fourier equation gives the outflow of heat shown in the figure. The difference between the heat outflow and inflow is the heat accumulated in the x-direction in the micro-hexahedron, that is,

$$-\lambda \frac{d\theta}{dx} dydzdt - \left[-\lambda \frac{d}{dx} \left(\theta + \frac{d\theta}{dx} dx \right) dydzdt \right] = \lambda \frac{d^2\theta}{dx^2} dxdydzdt.$$

The differential operators may be treated as ordinary fractions in calculations. This may be applicable to the y and z directions. In other words, the heat flux may be estimated in each of the three directions. How is the micro-hexahedron physically affected by this? The temperature change $d\theta$ occurs in the micro-hexahedron. The degree of this change depends on the material of the hexahedron; in short, it depends on whether the material is thermally heavy or light when the temperature is increased by means of thermal energy accumulated in the hexahedron. In terms of Newton's equations of motion, if the same force is applied to large and small material points, the point of large mass accelerates less than that of small mass, while the point of small mass is relatively easy to move. Such a property is represented by the heat capacity obtained by multiplying the volumetric specific heat by volume. Substituting this quantity into both sides of the above equation, we get

$$C_p\rho \cdot dxdydz \cdot d\theta = \lambda \left(\frac{d^2\theta}{dx^2} + \frac{d^2\theta}{dy^2} + \frac{d^2\theta}{dz^2} \right) dxdydzdt.$$

Rearrangement yields the three-dimensional unsteady-state heat transfer Eq. (2.3).

The unsteady-state heat transfer equation describes the physical phenomenon where heat flows by means of temperature difference. This temperature difference, which creates a driving force, is referred to as the potential difference. This equation may be principally applicable to a range of physical phenomena even if the potential difference and transferred object are different. For example, as discussed in Sect. 2.11, in moisture transfer, the potential difference is the difference in water vapour concentration. The *unsteady-state heat transfer equations* are often referred to as *diffusion equations*, because they describe physical phenomena where the potential difference is the driving force of heat or water vapour diffusion.

Fig. 2.3 Vibration of a
shearing stick

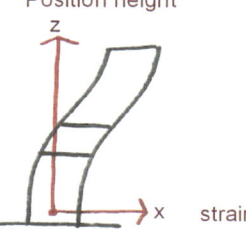

Position height

As mentioned in the footnote of page 4, the universal diffusion formulae are
referred to as Fick's laws (the first law is equivalent to Fourier's law of heat
diffusion and the second is the one-dimensional heat conduction equation).

A differential equation of the form (2.1), in which the first-order time derivative
equals a second-order space derivative expressed by a Laplacian (and perhaps other
constant coefficient advection terms) is classified as *parabolic*. Other classes of
partial differential equations are *hyperbolic* and *elliptic*. An example of the former
is the equation of vibration for a shearing stick, given by

$$\frac{\partial^2 x}{\partial t^2} = \frac{G}{\rho} \frac{\partial^2 x}{\partial z^2}, \tag{2.7}$$

where x and z are displacement [m] and height [m], respectively, in the coordinate
system of Fig. 2.3; G [N/m^2] is the modulus of elasticity; and ρ [kg/m^3] is the density
of shear stick. Note that the time derivative is second-order. Leading elliptic exam-
ples are the Poisson equation $\frac{\partial^2 \phi}{\partial x^2} + \frac{\partial^2 \phi}{\partial y^2} + \frac{\partial^2 \phi}{\partial z^2} + g = 0$ and the Laplace equation
$\frac{\partial^2 \phi}{\partial x^2} + \frac{\partial^2 \phi}{\partial y^2} + \frac{\partial^2 \phi}{\partial z^2} = 0$. These are equivalent to three-dimensional steady-state heat
conduction Eq. (2.3) (in which the time differential on the left side of the equation
is set to zero).

2.2   What is Discretization

The one-dimensional unsteady-state heat transfer equation, Eq. (2.1), can be defined
if the initial conditions and boundary conditions are known, but may not be analytically solvable. Solutions must then be sought numerically, which is the main theme
of this book. As already described above, continuous differential equations must
have been *discretized* before they can be solved by a computer. Put simply,
discretization is the replacement of infinitesimal time dt and distance dx with a finite
time Δt and distance Δx, respectively. These operations are known as *time* and *space*
discretizations, respectively. Of course, appropriate procedures must be followed in
the discretization process. Three basic methods for space discretization are

- Finite Difference Method, FDM
- Control Volume Method, CDM
- Finite Element Method, FEM

The finite difference and CVMs give the same discretization equations. However, in the former approach, a Taylor expansion is applied to the original differential equation, while the latter builds the balance between flux inputs and outputs into the control volume in order to satisfy the original differential equation. The latter, which is much more physically intuitive and easier to understand, is used as the basis of space discretization throughout this book. On the other hand, time discretization methods depend on the order of the time derivative in the original differential equation. In hyperbolic equations containing second order time differentials, Runge–Kutta method or similar methods, which enable multi-level integration, are used. First-order time derivatives are usually solved by one of the following methods:

- Forward FDM
- Crank–Nicolson method (central differences)
- Backward FDM

In principle, the difference scheme for time discretization is not limited to these methods and an infinite number of difference schemes can be created. This topic will be discussed later.

We emphasize that space and time discretization is inherently different and the two should be treated separately.

In practice, space discretization is applied to the original continuous system, followed by time discretization.

2.3 Space Discretization Based on Control Volume Method

As a representative example, let us consider a thermal field of semi-infinite soil, as shown in Fig. 2.4. The x-coordinate axis in the underground direction takes ground surface as its origin. Underground heat propagates only by conduction, but convective heat transfer occurs on the ground surface, which is exposed to external temperature θ_o [°C]. The convective heat transfer coefficient is α_o [W/(m^2 K)]. If the surface temperature is θ_1 [°C], the acquired heat is $\alpha_o(\theta_0 - \theta_1)$ [W/m^2]. On the ground surface, such phenomena occurs: a heat flux I [W/m^2] from known solar radiation (shortwave radiation), a loss due to latent heat of vaporization lE [W/m^2], a heat loss due to known long wave radiation flux R [W/m^2], and a latent heat of vaporization IE [W/m^2] that is the product of a latent heat of vaporization of water [J/kg] and a known evaporation [kg/(m^2 s)]. In summary, the problem amounts to analyze the evolution of the semi-infinite ground thermal field when thermal impact was applied only on the ground surface.

First, control-volume space discretization allocates the system to control volumes of limited size. The heat capacities of these control volumes $C_p \rho \Delta x$ [J/K] are represented by temperature *nodes* (since this is a one-dimensional problem, the area is implicitly assumed to be 1 m^2). As implied by this description, the temperature nodes are placed at the centers of the control volumes. This is referred to as lumped parameterization. Figure 2.4 illustrates five control volumes extending from the

Fig. 2.4 Space
discretization model based
on CVM in which the
surface layers of the semi-
infinite soil are lumped
parameterized

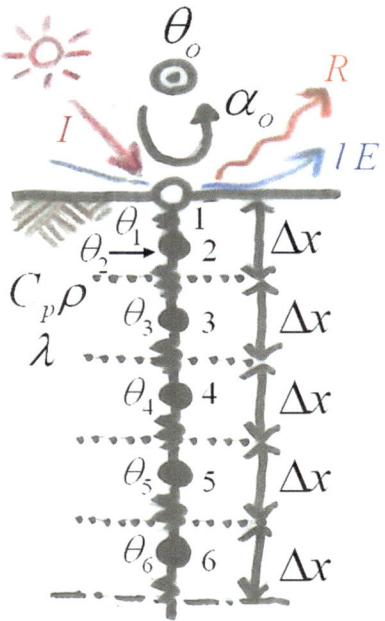

ground surface. To simplify the discussion, these volumes have the same thickness,
Δx. Ordinarily, near the ground where boundary conditions have a strong effect, the
allocated widths are small, making the intervals nonuniform. Another important
point of the problem is the degree of depth to be considered. Because thermal
impact is applied only to the semi-infinite one side of the boundary, the impact
qualitatively exerts from the ground surface downward a valid depth, suggesting
that the temperature beyond the depth is expected to stabilize. At this depth, the
temperature should be the average value for the thermal field determined by the
surface thermal impact, because no other thermal generation or absorption occurs
throughout the system. Indeed, if underground temperature is measured at depths
exceeding some critical depth, a constant temperature is reached, known as the
isothermal layer temperature. Therefore in this example, analysis must be
conducted until the depth reaches the point with the isothermal layer temperature.
This required degree of depth depends on the thermophysical properties of the
ground, i.e., the ease of heat transfer and the thermal mass $C_p\rho$, but approximately
10 m is sufficient. The temperatures at the lumped parameterized nodes in the
control volume are denoted from θ_2 to θ_6[°C]. These are unknown values to be
solved. An unknown temperature node is also placed on the ground surface; this
node is expressed as θ_1[°C]. Note that this node has no heat capacity. Nodes with
and without heat capacity are denoted as ● and ○, respectively. Moreover, the
temperatures at nodes designated ⊙ (in this case the external temperature) are
predefined; these nodes determine the boundary conditions stipulated by the heat
flux through convective heat transfer. Hence, they are referred to as (temperature)
stipulation nodes. We are now ready to program the simulation.

Heat balance equations will be formulated for temperature nodes 1–6. The right side of a heat balance equation should account for all heat flux elements flowing into and out of the control volume. The left side describes the resulting physical changes. Physical phenomena, as you already known, will occur (A temperature difference occurs in the control volume. Strictly speaking, it is balanced with the temperature change over time for the control volume with heat capacity.). The surface temperature nodes are subject not only to conductive heat but also to other various heat flows permitted by the boundary conditions.

At node 1, we have

$$0 = C_{12}(\theta_2 - \theta_1) + \alpha_o(\theta_o - \theta_1) + I - R - lE. \tag{2.8}$$

The first term on the right is the flux entering node 1 from node 2, and C_{12} is the heat conductance [W/(m^2 K)], defined as

$$C_{12} = \frac{\lambda}{\Delta x/2}. \tag{2.9}$$

The denominator is divided by the distance between nodes 1 and 2. The second term on the right of Eq. (2.8) indicates the flux flowing into node 1 through convection. Note that C_{12} and convection thermal conductivity have the same dimension of conductance. Moreover, in both terms, θ_1 is deducted from the adjoining temperature, because the inflow has a positive number. Similarly, the quantities R and lE are assumed to have positive influx. Moreover, because this problem is set up in one dimension, the area (1 m^2) is disregarded in both sides. Strictly, the equations should account for all the heat flux flowing into an element (as seen in the case where the heat flux entering the micro-hexahedron in the unsteady-heat conduction equation); multiplication by the area is required, but this is omitted for simplification. Finally, the left side of Eq. (2.8) is evaluated to be zero because the surface temperature nodes have no heat capacity (and no volume). In the system schematic shown in Fig. 2.4, the right side of the heat balance equation is always evaluated to be zero at the nodes indicated as ○.

Node 2 satisfies the following equations:

$$C_p\rho\Delta x\frac{d\theta_2}{dt} = C_{21}(\theta_1 - \theta_2) + C_{23}(\theta_3 - \theta_2), \tag{2.10}$$

$$C_{21} = \frac{\lambda}{\Delta x/2}, \tag{2.11}$$

$$C_{23} = \frac{\lambda}{\Delta x}, \tag{2.12}$$

where $C_{21} = C_{12}$. As should be physically understood, the conductance is symmetric and $C_{ij} = C_{ji}$. Here we describe the configuration of C_{21} assuming that node

Fig. 2.5 Composite
conductance

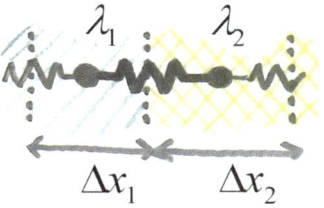

1 and node 2 have different heat conductivities and discretization widths. Let us
consider that the ground comprises multiple layers such as clay and gravel, as
shown in Fig. 2.5. In this case, *the composite conductance* is

$$C_{21} = \frac{1}{\frac{\Delta x_1/2}{\lambda_1} + \frac{\Delta x_2/2}{\lambda_2}}. \qquad (2.13)$$

If the thermal conductivity equals the discretization width, (2.13) reduces to
(2.12). Equation (2.13) is similar to the thermal transmittance, so-called U-value, an
important quantity in architectural environmental engineering. When the ground is
assumed to compose two layers of different conductance, it may behave like
tandem resistors in an electric circuit. For any materials with resistance, addition
might be performed. Whereas, conductivity ([W/(m K)] = [J/(s m K)]) of any
materials has thermal energy [J] in the numerator of its physical unit, implying
"ease of heat transfer." In this case, addition cannot be performed. First, the thermal
conductivity is divided by the thickness to obtain the conductance. Then, addition is
performed on the reciprocals of the values for conductance. After addition, the
reciprocal for the resulting sum is used for conductance. This general rule is
applicable not only to heat diffusion problems but also to the diffusion problems
involved with all types of potential fields.

The heat balance equations for nodes 3, 4, and 5 are as shown below. (The
conductance is obvious and not described explicitly.)

$$C_p \rho \Delta x \frac{d\theta_3}{dt} = C_{32}(\theta_2 - \theta_3) + C_{34}(\theta_4 - \theta_3), \qquad (2.14)$$

$$C_p \rho \Delta x \frac{d\theta_4}{dt} = C_{43}(\theta_3 - \theta_4) + C_{45}(\theta_5 - \theta_4), \qquad (2.15)$$

$$C_p \rho \Delta x \frac{d\theta_5}{dt} = C_{54}(\theta_4 - \theta_5) + C_{56}(\theta_6 - \theta_5). \qquad (2.16)$$

The temperature evolution for the final node (node 6) is

$$C_p \rho \Delta x \frac{d\theta_6}{dt} = C_{65}(\theta_5 - \theta_6). \qquad (2.17)$$

The right side contains a single term because all elements deeper than node
6 have the same temperature (namely, θ_6), and the conductive heat flux is zero.

Such a condition is referred to as an *adiabatic boundary*. Our thermal field of semi-infinite soil should ensure that the control volume lies at sufficient depth to assume the isothermal layer. The adiabatic boundary should be constructed at the bottom of this layer.

2.4 System State Equations

The heat balance equations for temperature nodes 1–6 are collectively expressed in a vector matrix Eq. (2.18). The matrix $\mathbf{M}$ contains the heat capacities $m_1, \cdots, m_6$ on the left side of the heat balance equation. In this example, $m_1 = 0$; for the remaining node i, $m_i = C_p \rho \Delta x$.

$$\mathbf{M} \frac{d\boldsymbol{\theta}}{dt} = \mathbf{C}\boldsymbol{\theta} + \mathbf{C_o}\boldsymbol{\theta_o} + \mathbf{f}. \tag{2.18}$$

Equation (2.18) is known as a *system state equation*. The system state equation collectively describes the balance equations. The matrix and vector elements are explicitly written below, but the reader must first substitute them into Eq. (2.18) for expansion and confirm that the elements match those in the heat balance equation for each node; Eqs. (2.8)–(2.10) and (2.14)–(2.17). This exercise is fundamental to understand numerical solutions of partial differential equations; hence, surely work on this exercise by yourself.

$$\boldsymbol{\theta} = \begin{bmatrix} \theta_1 \\ \theta_2 \\ \theta_3 \\ \theta_4 \\ \theta_5 \\ \theta_6 \end{bmatrix} = {}^T[\theta_1 \quad \cdots \quad \theta_6]. \tag{2.19}$$

$\boldsymbol{\theta}$ is a vector of unknown variables. The symbol T at the upper left of the vector (or matrix) indicates the *transpose*.[2] The transpose operator changes a horizontal (vertical) vector to a vertical (horizontal) one, and interchanges the row and column elements for matrices. We will encounter the transpose frequently in later chapters.

---

[2] Many textbooks show the transpose symbol at the upper right of the vector or matrix, but this convention may be easily confused with powers; hence, in this book, it is shown at the upper left.

$$\mathbf{f} = \begin{bmatrix} I - R - lE \\ \\ \\ \\ \\ \end{bmatrix}, \tag{2.20}$$

$$\mathbf{M} = \begin{bmatrix} m_1 & & & & & \\ & m_2 & & & & \\ & & m_3 & & & \\ & & & m_4 & & \\ & & & & m_5 & \\ & & & & & m_6 \end{bmatrix}. \tag{2.21}$$

Blank entries imply zero elements.

$$\mathbf{C} = \begin{bmatrix} -C_{12}-\alpha_o & C_{12} & & & & \\ C_{21} & -C_{21}-C_{23} & C_{23} & & & \\ & C_{32} & -C_{32}-C_{34} & C_{34} & & \\ & & C_{43} & -C_{43}-C_{45} & C_{45} & \\ & & & C_{54} & -C_{54}-C_{56} & C_{56} \\ & & & & C_{65} & -C_{65} \end{bmatrix}, \tag{2.22}$$

$$\mathbf{C_0} = \begin{bmatrix} \alpha_o \\ \\ \\ \\ \\ \end{bmatrix}, \tag{2.23}$$

$$\boldsymbol{\theta_0} = \begin{bmatrix} \theta_o{}^o \end{bmatrix}. \tag{2.24}$$

No doubt the reader has been surprised by the graceful quality of the system state equations.

For any parabolic diffusion equations, their system state equations are always expressed in the form of Eq. (2.18) after discretization. This means the state equations are universal formulation. In this example, the control volume method (CVM) is used for discretization. As known from the example, the system state equation takes the same formulation independent of the method of space discretization.

$\mathbf{M}$ is referred to as the *heat capacitance (capacity) matrix*. It has a regular structure and the heat capacities at the discretized elements are contained in its diagonal elements. The element (1,1) representing the ground surface is $m_1 = 0$. The values are also contained in diagonal elements in the case where the space finite difference method (FDM) is used for space discretization. There is a slight difference when calculus of finite element method (FEM) is used. Values are also contained in the non-diagonal elements. This interesting point will be explained in detail in later sections. For now, $\mathbf{M}$ can be assumed to be a diagonal matrix.

The vector matrix product $\mathbf{C_o \theta_o}$ indicates the boundary conditions of convective heat transfer as well as those determined by the stipulated temperature nodes. The vector $\mathbf{\theta_o}$ is a column vector of known temperature nodes; the stipulated temperature nodes. In this example, only the external temperature is treated; thus, the column vector $\mathbf{\theta_o}$ contains a single element. $\mathbf{C_o}$, where the number of rows is the number of unknown nodes and the number of columns is the number of aforementioned stipulated nodes, is a non-square matrix. If a heat relationship exists between the ith unknown number and the jth stipulated temperature node, the transport efficiency is contained in the (i,j)th element. In this example, because a convective heat flux occurs between the ground surface and the external environment, the convective heat transmittance α_o (its efficiency) is contained in element (1,1).

$\mathbf{C}$ is referred to as the *heat conductance matrix*. This matrix has a regular structure, and if a heat relationship exists between the ith and jth unknown temperature nodes, its transport efficiency is contained in the (i,j)th element. In this example, conductance $C_{12} = \frac{\lambda}{\Delta x/2}$ is contained in the element (1,2). Furthermore, this matrix has a fine symmetric structure. Only by treating the upper triangular elements, the lower triangle may be populated with the transposed values. This is an extremely useful characteristic in computer coding, but the symmetry is broken if a directional stipulation is imposed on the system, as happens with diodes in electrical circuits. This is applicable to the case where air is forcibly moved by a fan to transport heat together with an advection (see Sect. 2.9). At this point, it is assumed that $\mathbf{C}$ has a symmetric structure. The diagonal elements of $\mathbf{C}$ are sophisticated. The diagonal elements contain the row sums of the non-diagonal elements in the matrix $\mathbf{C}$ and the elements in $\mathbf{C_o}$, which is multiplied by -1. For example, element (1,1) in $\mathbf{C}$ contains a negative value of $(C_{12} + \alpha_o)$, which is the sum of the non-diagonal element in C and the element in the first row in $\mathbf{C_o}$. This relationship is again very useful in computer coding.

Once the system state equations are well known, it may be recognized that the heat balance equations, as have been described, do not need to be re-derived. This is the greatest advantage of system state equations. In other words, space discretization equations can be mechanically obtained without knowledge of the individual heat balance equations. This may present as a surprising fact.

The discretization procedure is summarized below:

- A vector of unknown variables is automatically determined when the block diagram (as in Fig. 2.4) is constructed.
- If space discretization is to be conducted via control volume or finite difference method, the heat capacitance matrix $\mathbf{M}$ contains the discretized heat capacities in its diagonal elements.
- Vector $\mathbf{\theta_o}$ of stipulated node temperatures is automatically determined.
- Matrix $\mathbf{C_o}$ containing the boundary conditions at the stipulated temperature nodes is constructed according to the following rules. If a heat relationship exists between the ith unknown temperature node and the jth stipulated temperature node, the conductance is contained in the element (i,j).
- The heat conductance matrix $\mathbf{C}$ is constructed according to the following rules. If a heat relationship exists between the ith and jth unknown temperature modes,

their conductance is contained in element (i,j). However, this operation is conducted solely on the upper triangular elements. The transposes may be copied into the lower triangular parts. The diagonal elements contain the row sums of the non-diagonal elements in $\mathbf{C}$ and the elements in $\mathbf{C_o}$, which is multiplied by -1.

Provided that the rules of the basic matrix structure are known, a given problem can always be turned into a system state equation similar to Eq. (2.18), and individual heat balance equations need not be solved. Standard matrix operations such as row summation and transpose are ideally suited to computer programs and are extremely valuable in creating a general-purpose program.

2.5 Time Discretization

The beauty and universality of system state equations has been widely appreciated, but Eq. (2.18) cannot yet be programmed into a computer. First, we must discretize the time derivative.

The left side of Eq. (2.18) is easily discretized as

$$\mathbf{M}\frac{d\boldsymbol{\theta}}{dt} = \frac{1}{\Delta t}\mathbf{M}(\boldsymbol{\theta}^{i+1} - \boldsymbol{\theta}^i). \tag{2.25}$$

The superscripted indices in the above equation are not exponentials, but represent the discretised time steps i and i + 1.

The right side of Eq. (2.18) is slightly problematic because we must decide at what point in time the vectors $\boldsymbol{\theta}$, $\boldsymbol{\theta_o}$, and $\mathbf{f}$ should be discretized; more specifically, whether they should be computed at the i th or $(i + 1)$ th time step. The former is a forward-difference computation; the latter constitutes backward difference. If the discretization is performed at the mid-points of both schemes, it becomes a Crank–Nicolson difference. In principle, any time point between the i th and $(i + 1)$ th steps is permitted, as already mentioned, an unlimited number of difference schemes is possible.

Let us consider the forward difference scheme in detail. If the state vector on the right side of (2.18) is computed at the i th step, the discretized equation is given as below:

$$\frac{1}{\Delta t}\mathbf{M}(\boldsymbol{\theta}^{i+1} - \boldsymbol{\theta}^i) = \mathbf{C}\boldsymbol{\theta}^i + \mathbf{C_o}\boldsymbol{\theta_o}^i + \mathbf{f}^i$$

$$\Leftrightarrow \boldsymbol{\theta}^{i+1} = \left[\frac{1}{\Delta t}\mathbf{M}\right]^{-1}\left\{\left[\frac{1}{\Delta t}\mathbf{M} + \mathbf{C}\right]\boldsymbol{\theta}^i + \mathbf{C_o}\boldsymbol{\theta_o}^i + \mathbf{f}^i\right\} \tag{2.26}$$

Similarly, if the state vector on the right side of (2.18) is computed at the $(i + 1)$ th step (backward-difference scheme), the equation becomes

$$\boldsymbol{\theta}^{i+1} = \left[\frac{1}{\Delta t}\mathbf{M} - \mathbf{C}\right]^{-1} \left\{\left[\frac{1}{\Delta t}\mathbf{M}\right]\boldsymbol{\theta}^{i} + \mathbf{C_o}\boldsymbol{\theta_o}^{i+1} + \mathbf{f}^{i+1}\right\}. \qquad (2.27)$$

The one-dimensional unsteady-state heat transfer Eq. (2.1) will generate complete numerical solutions. In Eqs. (2.26) and (2.27), all the known variables are contained on the right side. Note that the vectors $\boldsymbol{\theta_o}$ and $\mathbf{f}$ are referenced from archived data such as climate data, and so are independent of time step. Thus, the unknown variable vector $\boldsymbol{\theta}$ can be determined. In the first time step, $\boldsymbol{\theta}^1$ is computed from a vector of initial conditions. $\boldsymbol{\theta}^1$ is then substituted on the right side of the equation to obtain $\boldsymbol{\theta}^2$, and the procedure iterates. The initial condition vector must be specifically and appropriately declared; for example, the temperatures of all nodes can be set to 0 °C. In this manner, time integration is conducted through sequential calculation and a time series of unknown variable vectors is generated.

We have mentioned the need for care in deciding the initial conditions. If the initial conditions stray far from the solution, their effect will diminish only after numerous computations. To "appropriately declare" does not mean to arbitrarily declare, but means to precisely declare. When the heat capacity is extremely large, as in our example, particular attention is required. If the initial temperature is set to 100 °C at all nodes, for instance, and we wish to obtain the underground temperature profile for the period from December 31st, 12 a.m. to January 1st, 1 a. m., then the node temperatures will be discontinuous across this time-boundary. This situation does not allow us to stop the numerical calculation at the end of that particular year, thus we must continue one more year (or much more years) so as to obtain a smooth profile. In this manner, extremely inappropriate initial conditions exert a drastic, irreversible effect on the thermal system. Ground thermal conduction calculations are often performed annually on the basis of the same climate conditions. Ideally, the calculation time can be drastically reduced by making preliminary temperature estimates of the isothermal layer, which can be used as initial conditions. The above process is referred to as *annual steady calculations*.

Now let us revisit Eqs. (2.26) and (2.27). The index -1 at the upper right of the first term in both equations is the matrix inverse operator, which inverts the matrix. The inverse of a scalar is the reciprocal.[3] At this point, an intelligent reader would realize that by comparing Eqs. (2.26) and (2.27), the former calculation requires less effort. In the forward-difference scheme, the inverse of a diagonal matrix, namely $\left[\frac{1}{\Delta t}\mathbf{M}\right]$, is computed. The inverse of a diagonal matrix is the reciprocal of the diagonal elements (if a matrix is multiplied by its inverse, the identity matrix is produced),[4] precluding the need for sweep-out methods and other call routines that consume computational time. On the downside, the forward-difference scheme

---

[3] An inverse is any quantity that when multiplied by its original quantity, yields an identity. In scalars, the unit element is 1; in matrices, it is the unit matrix $\mathbf{E}$. If a unit is multiplied by a quantity, it yields the same quantity.

[4] No reverse can be defined for the element (1,1) at the Earth's surface temperature node because they are zero. For this reason, here explanation is made in general terms.

Fig. 2.6 Meaning of spatial
Crank–Nicolson method

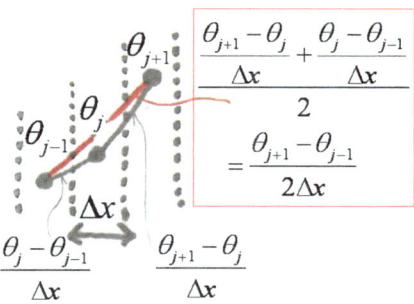

$$\frac{\dfrac{\theta_{j+1}-\theta_j}{\Delta x}+\dfrac{\theta_j-\theta_{j-1}}{\Delta x}}{2}$$

$$=\frac{\theta_{j+1}-\theta_{j-1}}{2\Delta x}$$

$$\frac{\theta_j-\theta_{j-1}}{\Delta x}\qquad\frac{\theta_{j+1}-\theta_j}{\Delta x}$$

is unstable unless a special condition is satisfied; more specifically, Δt cannot be set very large. This phenomenon can be heuristically understood as follows. The finite difference approximation is essentially a linear extrapolation from the current point. Hence, future predictions are likely to become unreliable if the time step Δt is large.

In contrast, in the backward-difference formulation, $\left[\frac{1}{\Delta t}\mathbf{M}-\mathbf{C}\right]$ becomes a band matrix with entries on both sides of the diagonal. In this case, the inverse matrix must be properly obtained through methods such as the sweep-out method, which are tedious to implement. However, unlike the forward difference formulation, a minimum Δt is not required to obtain a stable numerical solution. In the numerical analysis, although the size of the discretization error in the numerical solutions is of interest, far more important is whether the solutions diverge or stabilize. With respect to stability, even if the calculation requires effort, the backward difference formulation is recommended. The matrix $\mathbf{M}$ is diagonal when space discretization is performed using the control volume or finite difference methods. When the finite elements method is used, non-diagonal elements appear in $\mathbf{M}$ and the advantage of not requiring an inverse matrix calculation is lost.

We now introduce the concepts of *explicit* and *implicit* in classifying numerical methods. These concepts primarily define whether the inverse matrix must be solved when progressing from time steps i to $i+1$. In this sense, although forward-difference schemes have been called explicit, any scheme may be explicit or implicit depending on the space discretization method. To reiterate, if the FEM is used for space discretization, even if a forward-difference time discretization is adopted, the scheme is implicit rather than explicit.

Finally, we derive the Crank–Nicolson difference scheme. The Crank–Nicolson scheme is a central difference applied to spatial discretization, as shown in Fig. 2.6. The differential approximation at a discrete point j can be considered as the average spatial gradient between discrete points j and $j-1$, and j and $j+1$. The central difference gradient is the same between discrete points $j-1$ and $j+1$ because the space discretization is uniform; thus, the gradient at point j is the average of both gradients, as shown in the figure.

Applying the above ideas to the right side of Eq. (2.18), we obtain

$$
\frac{1}{\Delta t}\mathbf{M}(\boldsymbol{\theta}^{i+1} - \boldsymbol{\theta}^i) = \mathbf{C}\left\{\frac{1}{2}\boldsymbol{\theta}^i + \frac{1}{2}\boldsymbol{\theta}^{i+1}\right\} + \mathbf{C_0}\left\{\frac{1}{2}\boldsymbol{\theta_0}^i + \frac{1}{2}\boldsymbol{\theta_0}^{i+1}\right\} + \left\{\frac{1}{2}\mathbf{f}^i + \frac{1}{2}\mathbf{f}^{i+1}\right\}
$$

$$
\Leftrightarrow \boldsymbol{\theta}^{i+1} = \left[\frac{1}{\Delta t}\mathbf{M} - \frac{1}{2}\mathbf{C}\right]^{-1}
$$
$$
\left\{\left[\frac{1}{\Delta t}\mathbf{M} + \frac{1}{2}\mathbf{C}\right]\boldsymbol{\theta}^i + \mathbf{C_0}\left\{\frac{1}{2}\boldsymbol{\theta_0}^i + \frac{1}{2}\boldsymbol{\theta_0}^{i+1}\right\} + \left\{\frac{1}{2}\mathbf{f}^i + \frac{1}{2}\mathbf{f}^{i+1}\right\}\right\}.
$$

$$(2.28)$$

Note that the Crank–Nicolson difference is also an implicit method.

Equations (2.26)–(2.28) can be summarized and the forward, backward, and Crank–Nicolson differences are simultaneously expressed as

$$
\Leftrightarrow \boldsymbol{\theta}^{i+1} = \mathbf{A}^{-1}\left\{\mathbf{B}\boldsymbol{\theta}^i + \mathbf{C_0}\left\{(1-k)\boldsymbol{\theta_0}^i + k\boldsymbol{\theta_0}^{i+1}\right\} + \left\{(1-k)\mathbf{f}^i + k\mathbf{f}^{i+1}\right\}\right\}
$$
$$
\Leftrightarrow \boldsymbol{\theta}^{i+1} = \mathbf{A}^{-1}\mathbf{B}\boldsymbol{\theta}^i + \mathbf{A}^{-1}\left\{\mathbf{C_0}\left\{(1-k)\boldsymbol{\theta_0}^i + k\boldsymbol{\theta_0}^{i+1}\right\} + \left\{(1-k)\mathbf{f}^i + k\mathbf{f}^{i+1}\right\}\right\}
$$
$$
\Leftrightarrow \boldsymbol{\theta}^{i+1} = \mathbf{T}\boldsymbol{\theta}^i + (\text{heat impact on the system based on boundary conditions}).
$$

$$(2.29)$$

Forward, backward, and Crank–Nicolson differences are specified by $k = 0$, $k = 1/2$, and $k = 1$, respectively, although k can be assigned any arbitrary number $k \in [0,1]$.

2.6 Stability of Numerical Solutions

Observe the final Eq. (2.29) after time discretization of the system state equation. As already mentioned, the unknown variable vector in the next time step is influenced by both the variable vector of the current time step (that is not unknown anymore), and the boundary conditions, which is obtained through successive calculations (vector summation inside {} on the right side of Eq. (2.29)). Here we discuss the stability of this discretized system, ignoring the effect of the boundary conditions. Because the boundary conditions influence the discrete system from the outside, they are not relevant to the inherent stability of the system itself, i.e.,

Hence, the true impact of the aforementioned system is expressed as $\mathbf{T} = \left[\frac{1}{\Delta t}\mathbf{M} - k\mathbf{C}\right]^{-1}\left[\frac{1}{\Delta t}\mathbf{M} + (1-k)\mathbf{C}\right] \equiv \mathbf{A}^{-1}\mathbf{B}$. This matrix $\mathbf{T} \equiv \mathbf{A}^{-1}\mathbf{B}$ is a *transition matrix*, so-called because it embodies the characteristics of the time transition. If the second term on the right side in row 3 of the above equations is ignored, $\boldsymbol{\theta}^{i+1} = \mathbf{T}\boldsymbol{\theta}^i$, equivalent to geometric progression in scalar recursions. We now ask: what is the necessary and sufficient condition for convergence and stability of the general terms in the following geometric progression?

Fig. 2.7 Heat conduction inside walls

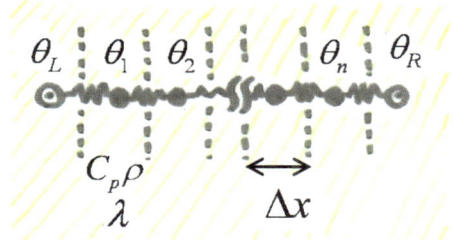

$$\{a_1, a_2, a_3, \ldots, a_n\} = \{a, ar, ar^2, \ldots, ar^{n-1}\} \Leftrightarrow a_n = r \cdot a_{n-1}$$

Here knowledge from junior high school may be useful, that is, a series converges if its geometric ratio r satisfies $|r| \leq 1$. The same idea applies to vector matrix recurrence formulae. However, the problem of how to measure the size of the transition matrix $\mathbf{T}$ arises. The answer lies in the *eigenvalues* of $\mathbf{T}$. Generally, an $n \times n$ square matrix has n eigenvalues. For convergence, it could be argued that the absolute value for the maximum eigenvalue should not exceed 1. In other words,[5]

$$|\text{Max}[\text{eigen}[\mathbf{T}]]| \leq 1. \tag{2.30}$$

We now apply forward-difference time discretization to the one-dimensional unsteady heat transfer system contained inside walls, as shown in Fig. 2.7. In this system, the temperatures inside the walls on both sides are specified at temperature nodes θ_L and θ_R. The wall between the boundary nodes is divided into n partitions. As before, space is discretized via the CVM. Following the basic rules of vectors and matrices described earlier, the state equations of this system (2.18) contains the following elements:

$$\boldsymbol{\theta} = \begin{bmatrix} \theta_1 \\ \vdots \\ \theta_n \end{bmatrix}, \tag{2.31.1}$$

$$\mathbf{f} = \mathbf{0}, \tag{2.31.2}$$

---

[5] This argument derives from the fact that the time evolution of the error between the numerical solution and the explicit solution obeys the original equation.

$$\mathbf{M} = \begin{bmatrix} C_p\rho \cdot \Delta x & & \\ & \ddots & \\ & & C_p\rho \cdot \Delta x \end{bmatrix}, \tag{2.31.3}$$

$$\mathbf{C} = \begin{bmatrix} -\dfrac{2\lambda}{\Delta x} & \dfrac{\lambda}{\Delta x} & & & \\ \dfrac{\lambda}{\Delta x} & -\dfrac{2\lambda}{\Delta x} & \dfrac{\lambda}{\Delta x} & & \\ & \ddots & \dfrac{\lambda}{\Delta x} & -\dfrac{2\lambda}{\Delta x} & \dfrac{\lambda}{\Delta x} \\ & & & \dfrac{\lambda}{\Delta x} & -\dfrac{2\lambda}{\Delta x} \end{bmatrix}, \tag{2.31.4}$$

$$\mathbf{C_o} = \begin{bmatrix} \dfrac{\lambda}{\Delta x} & 0 \\ 0 & 0 \\ \vdots & \vdots \\ 0 & 0 \\ 0 & \dfrac{\lambda}{\Delta x} \end{bmatrix}, \tag{2.31.5}$$

$$\boldsymbol{\theta_o} = \begin{bmatrix} \theta_L \\ \theta_R \end{bmatrix}. \tag{2.31.6}$$

In this situation, no heat sources (source; generation) or sinks (intake) exist in the walls, and so the vector $\mathbf{f}$ vanishes. The stipulated node temperature vector $\boldsymbol{\theta_o}$ is a column vector of length 2.

Explicitly writing the aforementioned matrix elements, the transition matrix of forward difference time discretization is obtained as

$$
\mathbf{T} = \mathbf{A}^{-1}\mathbf{B} =
\begin{bmatrix}
\frac{C_p\rho\cdot\Delta x}{\Delta t} & & \\
& \ddots & \\
& & \frac{C_p\rho\cdot\Delta x}{\Delta t}
\end{bmatrix}^{-1}
\begin{bmatrix}
\frac{C_p\rho\cdot\Delta x}{\Delta t} - \frac{2\lambda}{\Delta x} & \frac{\lambda}{\Delta x} & & & \\
\frac{\lambda}{\Delta x} & \frac{C_p\rho\cdot\Delta x}{\Delta t} - \frac{2\lambda}{\Delta x}\frac{\lambda}{\Delta x} & & & \\
& \ddots & \ddots & \ddots & \\
& & \frac{\lambda}{\Delta x} & \frac{C_p\rho\cdot\Delta x}{\Delta t} - \frac{2\lambda}{\Delta x} & \frac{\lambda}{\Delta x} \\
& & & \frac{\lambda}{\Delta x} & \frac{C_p\rho\cdot\Delta x}{\Delta t} - \frac{2\lambda}{\Delta x}
\end{bmatrix}
$$

$$
=
\begin{bmatrix}
1 - \frac{2\lambda\Delta t}{C_p\rho\Delta x^2} & \frac{\lambda\Delta t}{C_p\rho\Delta x^2} & & & \\
\frac{\lambda\Delta t}{C_p\rho\Delta x^2} & 1 - \frac{2\lambda\Delta t}{C_p\rho\Delta x^2}\frac{\lambda\Delta t}{C_p\rho\Delta x^2} & & & \\
& \ddots & \ddots & \ddots & \\
& & \frac{\lambda\Delta t}{C_p\rho\Delta x^2}1 - \frac{2\lambda\Delta t}{C_p\rho\Delta x^2} & \frac{\lambda\Delta t}{C_p\rho\Delta x^2} & \\
& & & \frac{\lambda\Delta t}{C_p\rho\Delta x^2} & 1 - \frac{2\lambda\Delta t}{C_p\rho\Delta x^2}
\end{bmatrix}
$$

$$
=
\begin{bmatrix}
1-2r & r & & & \\
r & 1-2r & r & & \\
& \ddots & \ddots & \ddots & \\
& & r & 1-2r & r \\
& & & r & 1-2r
\end{bmatrix}
=
\begin{bmatrix}
1 & & \\
& \ddots & \\
& & 1
\end{bmatrix}
+ r
\begin{bmatrix}
-2 & 1 & & & \\
1 & -2 & 1 & & \\
& \ddots & \ddots & \ddots & \\
& & 1 & -2 & 1 \\
& & & 1 & -2
\end{bmatrix}
= \mathbf{E} + r\mathbf{F},
$$

$$\tag{2.32}$$

where $r = \frac{\lambda\Delta t}{C_p\rho\Delta x^2}$ and $\mathbf{E}$ is the identity matrix. The eigenvalues of $\mathbf{F}$, which constitute an $n \times n$ square band matrix, are $-4\sin^2\left[\frac{t\pi}{2n}\right]$ (where $t = 1, 2, ..., n$). Given that when the eigenvalues for a matrix $\mathbf{D}$ are λ_D, the eigenvalues for a function of $\mathbf{D}$, $f(\mathbf{D})$, are $f(\lambda_D)$, the eigenvalues for the transition matrix are

$$
\lambda_t = 1 - 4r\sin^2\left[\frac{t\pi}{2n}\right] \quad \text{where } t = 1, 2, \cdots, n. \tag{2.33}
$$

If Eq. (2.30) is satisfied, $\left|1 - 4r\sin^2\left[\frac{t\pi}{2n}\right]\right| \leq 1$ is also satisfied. Hence, the *stability condition* for the forward-difference time discretization of this problem is obtained as

Fig. 2.8 Transition matrix
eigenvalues for backward
difference

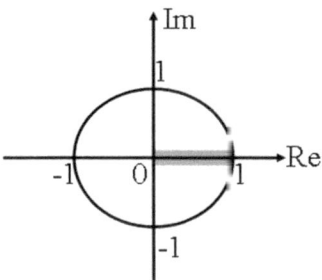

$$\Delta t \leq \frac{C_p \rho \Delta x^2}{2\lambda}.\qquad(2.34)$$

We now establish the stability condition of the backward-difference time
discretization and the Crank–Nicolson scheme using.

First, the backward-difference formulation is given as

$$\mathbf{T} = \mathbf{A}^{-1}\mathbf{B} = \left[\left[\frac{1}{\Delta t}\mathbf{M}\right]^{-1}\left[\frac{1}{\Delta t}\mathbf{M} - \mathbf{C}\right]\right]^{-1} = [\mathbf{E} - r\mathbf{F}]^{-1},\qquad(2.35)$$

and the eigenvalues for the transition matrix is written as

$$\lambda_t = \frac{1}{1 + 4r\sin^2\left[\frac{t\pi}{2n}\right]}.\qquad(2.36)$$

Recall that the inverse of a matrix is equivalent to the reciprocal of a scalar (see
page 14 footnote). Equation (2.36) always satisfies $0 \leq \lambda_t \leq 1$. In other words, if
backward difference is applied to time discretization, the numerical solutions are
unconditionally stable and Δt can be arbitrarily large. Furthermore, this result, i.e.,
that the eigenvalues are always positive and less than one, means that their absolute
values never exceed 1. Matrix eigenvalues are generally expressed as complex
numbers, such that if Eq. (2.36) is drawn on a Gaussian complex plane, Fig. 2.8 is
obtained. If none of the eigenvalues for a transition matrix exceed 1 they are
contained within the unit circle on the complex plane. In the backward-difference
formulation, the eigenvalues for the transition matrix lie along the positive real
number axis as shown in the figure. The lack of negative eigenvalues is the main
difference between this scheme and the Crank–Nicolson scheme. The next section
will elaborate on this fact, but here we mention that because Crank–Nicholson
difference permits negative eigenvalues, it induces "fluctuating numerical solutions," which are not physically possible, while not diverging. Therefore, the
preferred difference scheme is backward difference. This scheme guarantees
unconditional stability without numerical fluctuations.

Fig. 2.9 Transition matrix
eigenvalues for
Crank–Nicolson difference

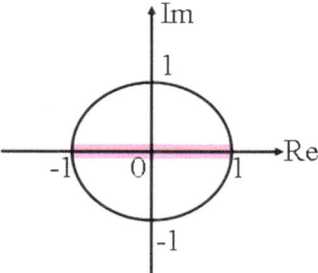

In the Crank–Nicolson difference scheme, we have

$$
\begin{aligned}
\mathbf{T} = \mathbf{A}^{-1}\mathbf{B} &= \left[\frac{1}{\Delta t}\mathbf{M} - \frac{1}{2}C \right]^{-1} \left[\frac{1}{\Delta t}\mathbf{M} + \frac{1}{2}C \right] \\
&= \left[\frac{2}{\Delta t}\mathbf{M} - C \right]^{-1} \left[\frac{2}{\Delta t}\mathbf{M} + C \right] \\
&= \left[\left[\frac{1}{\Delta t}\mathbf{M} \right]^{-1} \left[\frac{2}{\Delta t}\mathbf{M} - C \right] \right]^{-1} \left[\frac{1}{\Delta t}\mathbf{M} \right]^{-1} \left[\frac{2}{\Delta t}\mathbf{M} + C \right] \\
&= [2\mathbf{E} - r\mathbf{F}]^{-1}[2\mathbf{E} + r\mathbf{F}],
\end{aligned}
\tag{2.37}
$$

with the following eigenvalues for the transition matrix:

$$
\lambda_t = \frac{2 - 4r\sin^2\left[\frac{t\pi}{2n}\right]}{2 + 4r\sin^2\left[\frac{t\pi}{2n}\right]}.
\tag{2.38}
$$

Equation (2.38) always satisfies $-1 \le \lambda_t \le 1$. In other words, as with backward difference, the Crank–Nicolson is stable, regardless of the value of Δt, and the numerical solution always converges. If Eq. (2.38) is drawn in a Gaussian complex plane, Fig. 2.9 is obtained, in which the eigenvalues for the transition matrix lie within the range $[-1, 1]$. The presence of negative eigenvalues causes temporal and spatial fluctuations in the numerical solutions, as discussed in the next section.

2.7   Fluctuations in the Numerical Solutions

Basically, if k in the time-discretized system state equation (expression (2.29)) exceeds 1/2, Eq. (2.30) is satisfied, i.e., convergence of the numerical solutions is guaranteed.

However, as mentioned above, the negative eigenvalues for the transition matrix $\mathbf{T}$ introduce unwanted oscillations in the numerical solutions.

Fig. 2.10 Example of
temperature distribution
analysis in single wall;
(*above*) is initial state,
(*bottom*) is the steady-state

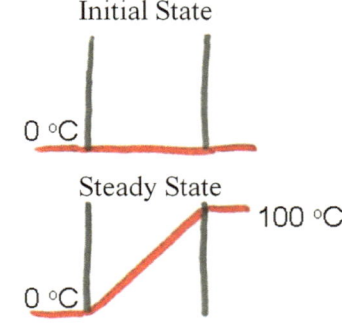

Fig. 2.11 Fluctuation
of numerical solutions

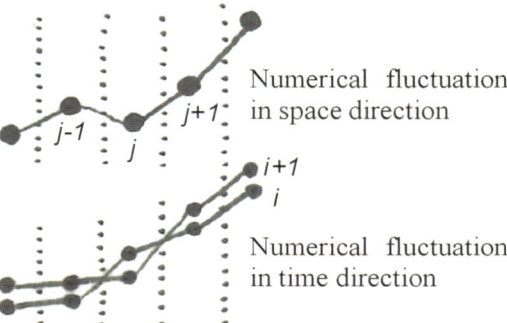

In the Crank–Nicolson difference scheme, we proved that the convergence condition (2.30) is always satisfied. Therefore, if we desire to less space-discretized error as discussed in Fig. 2.6, we may think the Crank–Nicolson difference scheme is better than the backward-difference scheme. However, this is not always the case. The author's experience teaches that the backward-difference scheme is best.

Consider one-dimensional unsteady heat transfer in which the temperature of a single layer wall is maintained at 0 °C at the left wall, but where the right wall is restricted to 100 °C (see Fig. 2.10). The upper and lower panels in the figure show the initial condition and the steady state temperature distribution after the passage of unlimited time, respectively. What happens if the Crank–Nicolson difference scheme is applied to time discretization? Of course, as explained in the previous section, stability is guaranteed, and so the numerical solutions will never diverge at any time. However, because negative eigenvalues for the transition matrix **T** exist, a slightly inconvenient situation may arise. For example, in Fig. 2.10, the temperature at a given point should always exceed that of its left neighbor. Physically, because the left wall is retained at 0 °C from $t = 0$, heat flows from the right side, where the temperature is restricted to 100 °C; thus, the temperature distribution should unambiguously increase from left to right. However if Crank–Nicolson difference is used, fluctuations such as those shown in the upper panel of Fig. 2.11 may occur. This situation is termed *numerical fluctuation in space direction*. Moreover, in the temporal direction, the temperature at any given point must increase from the temperature at the same point in the previous time. However, the Crank–Nicolson formulation permits the fluctuations shown in the lower panel in Fig. 2.11, termed *numerical fluctuation in time direction*. Again, negative eigenvalues for the transition matrix are responsible for these fluctuations, but no divergence occurs unless the absolute value for the maximum eigenvalue exceeds 1. That is, the numerical fluctuations are transient and as the system reaches steady-state, the numerical solutions approach the steady-state temperature distribution. Therefore, while not severely problematic, this situation should be avoided as the numerical solutions are transiently physically impossible.

When obtaining numerical solutions, the magnitude of errors in space and time discretization is a significant issue. Within the permitted discretization errors, it is natural to desire that time discretization width Δt is as large as possible and that time integration is maximally efficient. While the error magnitudes are certainly important, much more important is whether the numerical solutions stabilize and are physically plausible. Numerical inaccuracies in large jobs worth more than dozens thousands yen are especially undesirable, when you use a super-computer system, for example. From this viewpoint, the discretization method should be explored cautiously and time discretization should be formulated as backward difference, which is both stable and robust against fluctuations.

2.8   von Neumann Stability Analysis

The above analysis on the convergence and stability of numerical solutions is equivalent to the geometric progression of a scalar time-discretized equation, and builds on an intuitive understanding of the convergence conditions. Less intuitive but more mathematically rigorous is *von Neumann stability analysis*, discussed next.

The steps involved in this top-down approach are shown below. We suppose that the discretization equations in space and time directions are provided. The variable is ϕ.

1. Variable ϕ is time- and space-indexed by a right superscript n and a right subscript i, respectively, and is represented as ϕ_i^n.
2. All discretization variables are *discrete Fourier transformed* in the space direction as follows:

$$\phi_i^n = V^n \exp[Ik(i \cdot \Delta x)] = V^n \exp[Ii\varpi], \tag{2.39}$$

   where I is imaginary unit (expressed in uppercase to distinguish it from index i), k is the frequency, and ϖ is the phase angle ($k \cdot \Delta x$).

   For your reference, the discrete Fourier transform in the time and space directions is,

$$\phi_i^n = V^n \exp\left[I\left(k_{space}i\Delta x - k_{time}n\Delta t\right)\right]. \tag{2.40}$$

3. The discrete Fourier-transformed discretized equation is algebraically solved and rearranged as $V^{n+1} = G \cdot V^n$. Here G is called the *amplification coefficient*.
4. The stability condition of the discretized equation for a selected phase angle is

$$|G| \le 1 \text{ or } G \cdot \overline{G} \le 1, \tag{2.41}$$

where $\overline{G}$ is the complex conjugate of G.

Example Establish the stability of the numerical solutions of the advection equation $\frac{\partial \phi}{\partial t} + u \frac{\partial \phi}{\partial x} = 0$. Assume that the equation is space-discretized by the Crank–Nicolson method and time-discretized by backward difference.

Solution The scheme in which Crank–Nicolson and backward difference are applied to space and time discretization, respectively, is termed Backward Euler in Time Centered Space (BTSC).

The Crank–Nicolson spatial discretization of the advection equation is given as below (refer to Fig. 2.6 for a visual interpretation):

$$\frac{\phi_i^{n+1} - \phi_i^n}{\Delta t} + u \frac{\phi_{i+1}^{n+1} - \phi_{i-1}^{n+1}}{2\Delta x} = 0.$$

The discrete Fourier transforms of the φ variables in this equation are $\phi_i^n = V^n \exp[Ii\varpi]$, $\phi_i^{n+1} = V^{n+1} \exp[Ii\varpi]$, $\varphi_{i+1}^{n+1} = V^{n+1} \exp[I(i+1)\varpi]$, and $\phi_{i-1}^{n+1} = V^{n+1} \exp[I(i-1)\varpi]$.

Substituting these into the discretized equation, we obtain

$$V^{n+1}\exp[Ii\varpi] - V^n\exp[Ii\varpi] + \frac{u\Delta t}{2\Delta x}[V^{n+1}\exp[I(i+1)\varpi] - V^{n+1}\exp[I(i-1)\varpi]] = 0.$$

In terms of the *Courant number*, $C = \frac{u\Delta t}{\Delta x}$. After some rearrangement, this expression becomes

$$\left[1 + \frac{C}{2}(\exp[I\varpi] - \exp[-I\varpi])\right] V^{n+1} = V^n \Leftrightarrow G = \frac{1}{1 + IC \sin \varpi} = \frac{1 - IC \sin \varpi}{1 + C^2 \sin^2 \varpi}.$$

In converting the expression left of the second equals sign to that on the right, the denominator is made real via the *trigonometric relationships*:

$$\sin \theta \equiv \frac{\exp(I\theta) - \exp(-I\theta)}{2I}, \cos\theta \equiv \frac{\exp(I\theta) + \exp(-I\theta)}{2}, \sin^2\theta + \cos^2\theta = 1.$$

At this point, it is useful to recall the corresponding *hyperbolic functions*:

$$\sinh\theta \equiv \frac{\exp(\theta) - \exp(-\theta)}{2}, \cosh\theta \equiv \frac{\exp(\theta) + \exp(-\theta)}{2}, \cos h^2\theta - \sinh^2\theta = 1.$$

Hence, the following relationship is obtained:

$$|G| = \frac{\sqrt{1 + C^2 \sin^2 \varpi}}{1 + C^2 \sin^2 \varpi} = \frac{1}{\sqrt{1 + C^2 \sin^2 \varpi}} \le 1.$$

Fig. 2.12 G in a complex plane as per example

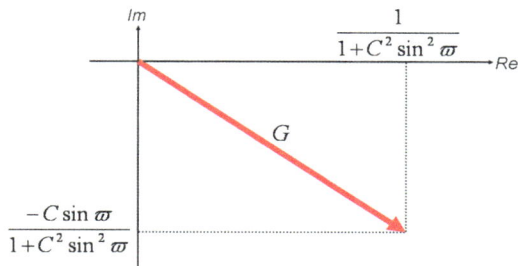

If the magnitude of the amplification coefficient G is drawn on a Gaussian complex plane, Fig. 2.12 is obtained.

The above analysis demonstrates that the BTSC discretization of the advection equation is unconditionally stable.

We now apply von Neumann stability analysis to the one-dimensional unsteady-state heat conduction Eq. (2.1). We assume that space is discretized by the CVM.

We first consider forward-difference time discretization. The discretization equation is

$$\frac{\theta_i^{n+1} - \theta_i^n}{\Delta t} = \frac{\lambda}{C_p\rho} \frac{\dfrac{\theta_{i+1}^n - \theta_i^n}{\Delta x} - \dfrac{\theta_i^n - \theta_{i-1}^n}{\Delta x}}{\Delta x}$$

$$\Leftrightarrow \frac{\theta_i^{n+1} - \theta_i^n}{\Delta t} = \frac{\lambda}{C_p\rho} \frac{\theta_{i+1}^n - 2\theta_i^n + \theta_{i-1}^n}{\Delta x^2}. \tag{2.42}$$

This formulation should be familiar to readers who have carefully studied Sect. 2.5. Applying the discrete Fourier transformation to (2.42) gives

$$\frac{1}{\Delta t}\left[V^{n+1}\exp(Ii\varpi) - V^n\exp(Ii\varpi)\right]$$

$$= \frac{\lambda}{C_p\rho\Delta x^2}\left[V^n\exp(I(i+1)\varpi) - 2V^n\exp(Ii\varpi) + V^n\exp(I(i-1)\varpi)\right]$$

$$\Leftrightarrow V^{n+1} = \left[1 + \frac{\lambda\Delta t}{C_p\rho\Delta x^2}[-2 + \exp(I\varpi) + \exp(-I\varpi)]\right]V^n, \tag{2.43}$$

with amplification coefficient

$$G = 1 - 2\frac{\lambda\Delta t}{C_p\rho\Delta x^2}(1 - \cos\varpi). \tag{2.44}$$

The stability condition that satisfies $|G| \leq 1$ is

$$\frac{\lambda\Delta t}{C_p\rho\Delta x^2} \leq \frac{1}{2}, \tag{2.45}$$

which clearly matches Eq. (2.34). To obtain a stable numerical solution, Δt cannot be arbitrarily large.

Next, we consider the case of backward-difference time discretization. The discretized equation is

$$\frac{\theta_i^{n+1} - \theta_i^n}{\Delta t} = \frac{\lambda}{C_p\rho} \frac{\theta_{i+1}^{n+1} - 2\theta_i^{n+1} + \theta_{i-1}^{n+1}}{\Delta x^2}. \tag{2.46}$$

Applying the discrete Fourier transform to (2.46) yields

$$\frac{1}{\Delta t}\left[V^{n+1}\exp(Ii\varpi) - V^n\exp(Ii\varpi)\right]$$

$$= \frac{\lambda}{C_p\rho\Delta x^2}\left[V^{n+1}\exp(I(i+1)\varpi) - 2V^{n+1}\exp(Ii\varpi) + V^{n+1}\exp(I(i-1)\varpi)\right]$$

$$\Leftrightarrow \left[1 - \frac{\lambda\Delta t}{C_p\rho\Delta x^2}[-2 + \exp(I\varpi) + \exp(-I\varpi)]\right]V^{n+1} = V^n. \tag{2.47}$$

Rearranging (2.47) as $V^{n+1} = G \cdot V^n$, the amplification coefficient is obtained as

$$G = \frac{1}{1 + 2\frac{\lambda\Delta t}{C_p\rho\Delta x^2}(1 - \cos\varpi)}. \tag{2.48}$$

Because the denominator clearly exceeds 1, $|G| \leq 1$ is always satisfied, and (2.42) is unconditionally stable.

Finally, let us apply the Crank–Nicolson difference scheme to time discretization. The discretization equation is

$$\frac{\theta_i^{n+1} - \theta_i^n}{\Delta t} = \frac{\lambda}{2C_p\rho\Delta x^2}\left[\theta_{i+1}^{n+1} + \theta_{i-1\theta}^{n+1} - 2\theta_i^{n+1} + \theta_{i+1}^n + \theta_{i-1}^n - 2\theta_i^n\right]. \tag{2.49}$$

Applying the discrete Fourier transform to (2.49) we get

$$\frac{1}{\Delta t}\left[V^{n+1}\exp(Ii\varpi) - V^n\exp(Ii\varpi)\right] = \frac{\lambda}{2C_p\rho\Delta x^2}\left[V^{n+1}\exp(I(i+1)\varpi) - 2V^{n+1}\exp(Ii\varpi)\right.$$

$$\left. + V^{n+1}\exp(I(i-1)\varpi) + V^n\exp(I(i+1)\varpi) - 2V^n\exp(Ii\varpi) + V^n\exp(I(i-1)\varpi)\right]$$

$$\Leftrightarrow \left[1 - \frac{\lambda\Delta t}{2C_p\rho\Delta x^2}[-2 + \exp(I\varpi) + \exp(-I\varpi)]\right]V^{n+1}$$

$$= V^n\left[1 + \frac{\lambda\Delta t}{2C_p\rho\Delta x^2}[-2 + \exp(I\varpi) + \exp(-I\varpi)]\right]. \tag{2.50}$$

Rearranging (2.50) and expressing as $V^{n+1} = G \cdot V^n$, the amplification coefficient is

$$G = \frac{1 - \frac{\lambda \Delta t}{C_p \rho \Delta x^2}(1 - \cos \varpi)}{1 + \frac{\lambda \Delta t}{C_p \rho \Delta x^2}(1 - \cos \varpi)}. \tag{2.51}$$

Evidently, the absolute value for the denominator exceeds that of the numerator and so $|G| \le 1$ is always satisfied. Although this system guarantees permanent stability, the amplification coefficient allows $\mathrm{Re} < 0$; thus, fluctuations may develop, as discussed in the previous section.

2.9 Heat System Applications

In this section, we describe three specific examples of heat transfer systems. In each case, the vectors and matrices of the spatially discretized system state equation are explicitly expressed.

Exercise 1 Consider a wall heat transfer problem in which convection heat transfer boundaries exist on both sides. Each wall is split into two sections and nodes with no heat capacity are set up on both surfaces.

Solution Recall the system state equation $\mathbf{M}\frac{d\boldsymbol{\theta}}{dt} = \mathbf{C}\boldsymbol{\theta} + \mathbf{C_o}\boldsymbol{\theta_o} + \mathbf{f}$.

On the basis of Fig. 2.13, the vector of unknown temperature nodes is $^{\mathrm{T}}\boldsymbol{\theta} = [\,\theta_1 \quad \theta_2 \quad \theta_3 \quad \theta_4\,]$, and the heat capacitance matrix is

$$\mathbf{M} = \begin{bmatrix} 0 & & & \\ & \dfrac{C_p \rho \cdot \ell}{2} & & \\ & & \dfrac{C_p \rho \cdot \ell}{2} & \\ & & & 0 \end{bmatrix}.$$

The vector–matrix product includes the boundary conditions imposed at the stipulated temperature nodes;

$$\mathbf{C_o} = \begin{bmatrix} \alpha_o & \\ & \\ & \alpha_r \end{bmatrix} \qquad \boldsymbol{\theta_o} = \begin{bmatrix} \theta_o \\ \theta_r \end{bmatrix}.$$

The heat flux boundary condition vectors are $\mathbf{f} = \mathbf{0}$.

Fig. 2.13 Heat system of
Exercise 1 (see text for
details)

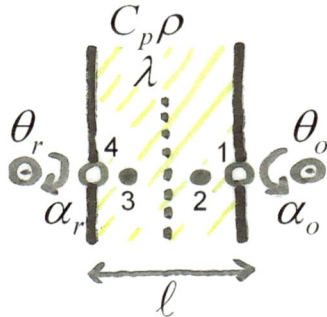

The heat conductance matrix is

$$
\mathbf{C} =
\begin{bmatrix}
-\dfrac{\lambda}{\ell/4} - \alpha_o & \dfrac{\lambda}{\ell/4} & & \\[2.2ex]
\dfrac{\lambda}{\ell/4} & -\dfrac{6\lambda}{\ell} & \dfrac{\lambda}{\ell/2} & \\[2.2ex]
& \dfrac{\lambda}{\ell/2} & -\dfrac{6\lambda}{\ell} & \dfrac{\lambda}{\ell/4} \\[2.2ex]
& & \dfrac{\lambda}{\ell/4} & -\dfrac{\lambda}{\ell/4} - \alpha_r
\end{bmatrix}.
$$

Exercise 2 Now consider a single-room model in which external air is introduced
by a fan. The airflow (i.e., ventilation) causes the air inside the room to be affected
by the external air temperature. The ventilation is Q [m^3/s]. The volume of the room
and the area of the wall are as shown in Fig. 2.14, along with the physical properties
of the wall. The volumetric specific heat is $(C_p\rho)_{air}$ [J/(m^3 K)].

(Hint) Construct separate heat balance equations for the room temperature (θ_r) at
node 6.

Solution The vector of unknown temperatures is $^T\boldsymbol{\theta} = \begin{bmatrix} \theta_1 & \theta_2 & \theta_3 & \theta_4 & \theta_5 & \theta_r \end{bmatrix}$,
and the vector of heat flux boundary conditions is $\mathbf{f} = \mathbf{0}$.

The heat capacitance matrix is

Fig. 2.14 Heat system
of Exercise 2 (see text
for details)

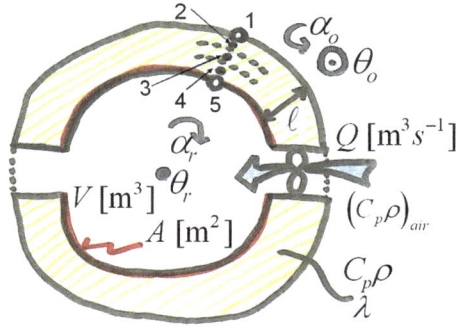

$$\mathbf{M} = \begin{bmatrix} 0 & & & & \\ & \dfrac{C_p \rho \ell A}{3} & & & \\ & & \dfrac{C_p \rho \ell A}{3} & & \\ & & & \dfrac{C_p \rho \ell A}{3} & \\ & & & & 0 \\ & & & & & V\left(C_p \rho\right)_{air} \end{bmatrix}.$$

   This problem deals simultaneously with the one-dimensional heat transfer in the
wall and the heat balance of the room; hence, the volume in the heat capacitance
matrix cannot be represented by the wall thickness alone (assuming a surface area
of 1 m^2) as before. Note that the discretized wall element and room volume are
clearly distinguished in the matrix.

   The vectors $\mathbf{C_0}$ and $\boldsymbol{\theta_0}$ expressing the boundary conditions at the stipulated
temperature nodes are

$$\mathbf{C_0} = \begin{bmatrix} \alpha_o A \\ 0 \\ 0 \\ 0 \\ 0 \\ Q\left(C_p \rho\right)_{air} \end{bmatrix}, \quad \boldsymbol{\theta_0} = [\theta_o].$$

$Q(C_p\rho)_{air}$ is the heat conductance due to the ventilation of the external air and room
temperature gradients. The unit is [m^3/s][J/(m^3 K)] = [W/K]. Note that each
element in the heat conductance matrix has the same dimensions (those of the
heat transfer coefficient multiplied by the surface area).

Fig. 2.15 Heat system of
Exercise 3 (see text for
details)

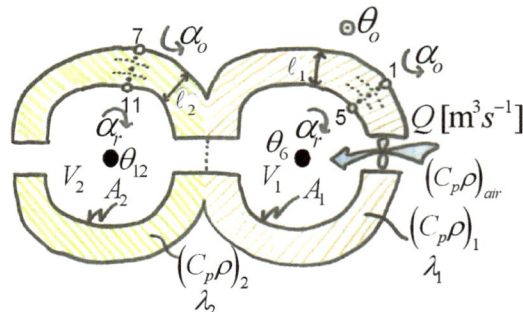

The heat conductance matrix is

$$
\mathbf{C} =
\begin{bmatrix}
-\dfrac{\lambda A}{\ell/6}-\alpha_o A & \dfrac{\lambda A}{\ell/6} & & & & \\[2mm]
\dfrac{\lambda A}{\ell/6} & -\dfrac{9\lambda A}{\ell} & \dfrac{\lambda A}{\ell/3} & & & \\[2mm]
& \dfrac{\lambda A}{\ell/3} & -\dfrac{6\lambda A}{\ell} & \dfrac{\lambda A}{\ell/3} & & \\[2mm]
& & \dfrac{\lambda A}{\ell/3} & -\dfrac{9\lambda A}{\ell} & \dfrac{\lambda A}{\ell/6} & \\[2mm]
& & & \dfrac{\lambda A}{\ell/6} & -\dfrac{\lambda A}{\ell/6}-\alpha_r A & \alpha_r A \\[2mm]
& & & & \alpha_r A & -\alpha_r A-Q\left(C_p\rho\right)_{air}
\end{bmatrix}.
$$

Again, the one-dimensional assumptions no longer apply, and the equations must account for the wall surface area. The reader should confirm that the dimensions of $\frac{\lambda A}{\ell/6}$ are those of $Q(C_p\rho)_{air}$, i.e., [W/(m K)][m^2]/[m] = [W/K]. The uncertain reader should substitute the aforementioned vector and matrix into the system state equation and rearrange to obtain the heat balance equation at each temperature node. In particular, it should be confirmed that the right side of the room's heat balance equation, which describes the change in room temperature, is driven by both ventilation from external air and heat convection with wall.

Exercise 3 Finally, consider two rooms connected in series ("Kamakura"; Japanese snow dome), as shown in Fig. 2.15. External air forced from a fan into room 1 enters room 2.

Solution Unknown temperature nodes are assigned in the following order: wall of room 1, room 1, wall of room 2, and room 2, as follows:

$$
{}^T\boldsymbol{\theta} = [\,\theta_1 \quad \cdots \quad \theta_5 \quad \theta_6 \quad \theta_7 \quad \cdots \quad \theta_{11} \quad \theta_{12}\,].
$$

The vector of heat flux boundary conditions is $\mathbf{f} = \mathbf{0}$.
The heat capacitance matrix is

$$
\mathbf{M} =
\begin{bmatrix}
0 & & & & & & & & \\
 & \dfrac{\left(C_p \rho^p\right)_1 \ell_1 A_1}{3} & & & & & & & \\
 & & \dfrac{\left(C_p \rho^p\right)_1 \ell_1 A_1}{3} & & & & & & \\
 & & & \dfrac{\left(C_p \rho^p\right)_1 \ell_1 A_1}{3} & & & & & \\
 & & & & 0 & & & & \\
 & & & & & V_1 \left(C_p \rho\right)_{air} & & & \\
 & & & & & & 0 & & \\
 & & & & & & & \dfrac{\left(C_p \rho^p\right)_2 \ell_2 A_2}{3} & \\
 & & & & & & & & \dfrac{\left(C_p \rho^p\right)_2 \ell_2 A_2}{3} \\
 & & & & & & & & \dfrac{\left(C_p \rho^p\right)_2 \ell_2 A_2}{3} \\
 & & & & & & & & 0 \\
 & & & & & & & & V_2 \left(C_p \rho\right)_{air}
\end{bmatrix}
$$

and the vectors $\mathbf{C_0}$ and $\boldsymbol{\theta_0}$ expressing the boundary conditions at the stipulated temperature nodes are

$$
{}^{\mathrm{T}}\mathbf{C_0} = \begin{bmatrix} \alpha_o A_1 & 0 & 0 & 0 & 0 & Q\left(C_p \rho\right)_{air} & \alpha_o A_2 & 0 & 0 & 0 & 0 & 0 \end{bmatrix}, \quad \boldsymbol{\theta_0} = [\theta_o],
$$

whose meaning should be clear from Example 2. However, the heat conductance matrix is more complicated than the previous example and is expressed as below (please see the following page).

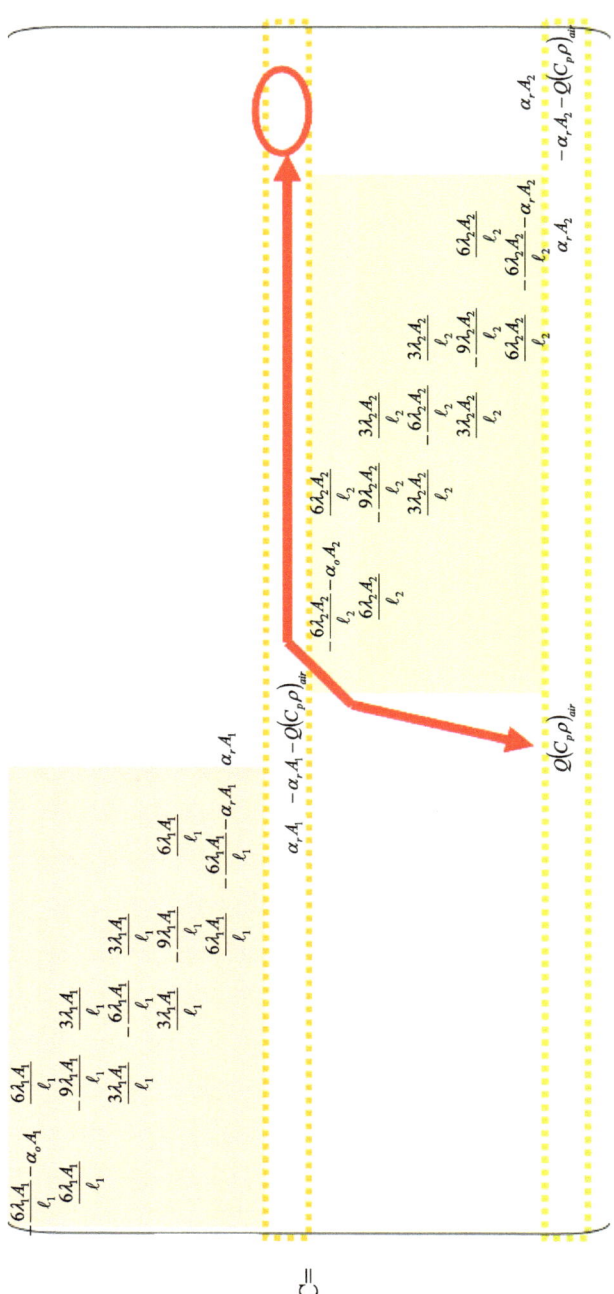

Because this is a large matrix (containing 12×12 elements), it is expressed as a figure rather than as a numerical equation. The sub-matrices describing heat conductance in rooms 1 and 2 are shaded orange and yellow, respectively, consistent with Fig. 2.15, in which the respective walls are indicated by the same colors. Rows 6 and 12 couple the heat balances between the two rooms. The most important contribution comes from element (12, 6). The 12th node (in room 2) is coupled to the 6th node (in room 1) via the ventilation conductance $Q(C_p\rho)_{air}$, but its symmetrical element (6,12), i.e., the heat conducted to the 6th node from node 12, is zero. This is understandable in terms of the heat balance equations of both rooms; room 2 receives ventilated air from room 1 blown in by the fan, whereas room 1 receives no equivalent airflow from room 2. This physical asymmetry renders the heat conductance matrix C non-symmetric. In this manner, when heat transfer is unidirectional (another example is a diode in an electrical circuit), C may be upper or lower triangular rather than symmetric.

2.10   Linearization of Radiant Heat Transfer

The *linearization of radiant heat transfer* covered in this section, while slightly off-topic, is important to understanding the symmetry of the conductance matrix and should thus be perused carefully.

Besides conduction and convection, heat transfer may occur by long-wave radiation (in contrast to visible light, which is a form of short-wave radiation; here the term "radiation" is restricted to long-wave radiation). Radiant heat transfer is considerably different from the conduction and convection previously discussed. Conductive and convective heat fluxes are proportional to the temperature difference between two points (conduction is described by the Fourier equation, while in convection, the heat transfer coefficient is multiplied by the temperature difference). Thermal radiation, on the other hand, is the propagation of electromagnetic waves that exert heat-like effects. The heat flux from environmental radiation [W/m^2] is expressed as

$$q_{rad} = \varepsilon \cdot \sigma \cdot T^4, \tag{2.52}$$

where ε is the emissivity, a dimensionless quantity equal to 1 for a perfectly black body and 0 for an ideal mirror. As implied by Eq. (2.52), ε is the efficiency of emission. It also denotes the absorption efficiency of arriving radiation, i.e., the absorptivity (Kirchhoff's law). σ is the Stephan–Boltzmann constant, equal to 5.67×10^{-8} [W/(m^2 K^4)], T is the surface temperature of the object [K]. The Stephan–Bolzmann constant can also be expressed in terms of the constant $C_b = 5.67$ [$10^{-8} \times$ W/(m^2 K^4)] to yield

$$q_{rad} = \varepsilon \cdot C_b \cdot \left(\frac{T}{100}\right)^4. \tag{2.53}$$

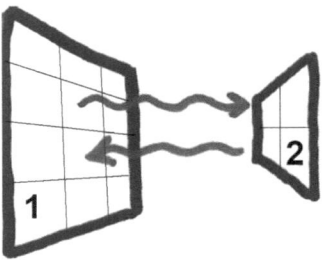

To obtain the net volume of radiant heat exchange between two surfaces (see
Fig. 2.16), we denote the temperature, emissivity, and area of surfaces 1 and 2 by T,
ε, and A, respectively. The surface to surface configuration factors when looking at
surface 2 from surface 1 and surface 1 from surface 2 are denoted as F_{12} and F_{21},
respectively. The surface configuration factor F_{ab} is the proportion of the target
surface b visible from the full field of vision seen by an "eye" placed at surface
a (if the "observing" surface is flat, the full field of vision is the hemisphere
covering the surface). In other words, lines of visions are projected from infinitesimal surface elements on the surface a on the basis of some rule (forming equi-solid
angles, not dissimilar to the spines on a hedgehog). The surface to surface form
factor is computed as the number of rays reaching surface b as a proportion of those
emitted from surface a, integrated over surface a. A detailed explanation of the
form factor is beyond the scope of this book; readers should refer to a standard
textbook of building physics or building environmental engineering. Readers
should also be familiar with reciprocity theorem and the important properties of
the form factor, introduced next.

Returning to our original theme, the net radiant heat exchange between surfaces
1 and 2, $H_{1 \rightarrow 2}$ [W], is expressed as

$$
\begin{aligned}
H_{1\rightarrow2} &= \varepsilon_2 F_{12} A_1 \varepsilon_1 \sigma T_1{}^4 - \varepsilon_1 F_{21} A_2 \varepsilon_2 \sigma T_2{}^4 \\
&= \varepsilon_1 \varepsilon_2 A_1 F_{12} C_b \left(\frac{T_1}{100}\right)^4 - \varepsilon_1 \varepsilon_2 A_2 F_{21} C_b \left(\frac{T_2}{100}\right)^4.
\end{aligned}
\tag{2.54}
$$

In the first term on the right side; emitted radiation, expressed as $\left(\varepsilon_1 C_b \left(\frac{T_1}{100}\right)^4 A_1\right)$,
eventually reaches surface 2 as $\left(\varepsilon_1 C_b \left(\frac{T_1}{100}\right)^4 A_1 F_{12}\right)$. The amount of heat absorbed by
surface 2 is $\left(\varepsilon_1 C_b \left(\frac{T_1}{100}\right)^4 A_1 F_{12} \varepsilon_2\right)$. The second term in (2.54) is the heat emitted from
surface 2, some of which is absorbed by surface 1. The difference between the two
transfers is the net amount of radiant heat exchange between the surfaces (from the
perspective of surface 1).

The reciprocity theorem of the surface to surface form factor (see Fig. 2.17)
states that

$$
A_1 F_{12} = A_2 F_{21}.
\tag{2.55}
$$

Fig. 2.17 Reciprocity
theorem applied to surfaces
1 and 2

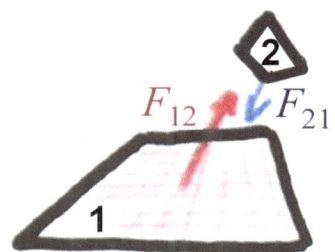

Applying (2.55) to Eq. (2.54), we get

$$H_{1 \to 2} = \varepsilon_1 \varepsilon_2 C_b A_1 F_{12} \left[\left(\frac{T_1}{100} \right)^4 - \left(\frac{T_2}{100} \right)^4 \right]$$

$$= A_1 F_{12} \cdot \varepsilon_1 \varepsilon_2 C_b \left[\frac{\left(\frac{T_1}{100} \right)^4 - \left(\frac{T_2}{100} \right)^4}{T_1 - T_2} \right] \cdot (\theta_1 - \theta_2). \tag{2.56}$$

Note that the absolute temperature difference is equal to the change in Celsius temperature. The term within the square brackets is known as the temperature coefficient. Provided that temperature differences are typical of ambient environments and do not exceed a few hundred °C, the temperature coefficient can be assumed as 1.

$$\frac{\left(\frac{T_1}{100} \right)^4 - \left(\frac{T_2}{100} \right)^4}{T_1 - T_2} \cong 0.04 \left[\frac{(T_1 + T_2)/2}{100} \right]^3 \cong 1 \tag{2.57}$$

Excluding lustrous surfaces such as metals and glasses, the emissivity is also close to 1 (typically 0.9).[6] Thus, the term $\varepsilon_1 \varepsilon_2 C_b \left[\left(\frac{T_1}{100} \right)^4 - \left(\frac{T_2}{100} \right)^4 \right] / [T_1 - T_2]$ in Eq. (2.56) can be treated as a constant. Substituting reasonable values into this term, we obtain $0.9 \times 0.9 \times 5.67 \times 1 = 4.6$. This value, which has units of conductance [W/(m² K)], is known as the radiant heat transfer rate, α_{rad}, defined as

$$\alpha_{rad} \equiv \varepsilon_1 \varepsilon_2 C_b \left[\frac{\left(\frac{T_1}{100} \right)^4 - \left(\frac{T_2}{100} \right)^4}{T_1 - T_2} \right] \cong 4.6 \left[W/(m^2 K) \right]. \tag{2.59}$$

---

[6] Solar radiation absorption varies greatly with color of the wall surfaces. Perfectly black bodies absorb all radiation (absorption = 1), while white surfaces can absorb as little as 0.5 of incoming radiation. However, emissivity is insensitive to surface color and is typically around 0.9.

In terms of the above linearized radiant transfer, Eq. (2.56) becomes

$$H_{1\to2} = A_1 F_{12} \cdot \alpha_{rad}(\theta_1 - \theta_2)$$
$$= -A_2 F_{21} \cdot \alpha_{rad}(\theta_2 - \theta_1) = -H_{2\to1}. \tag{2.60}$$

The format of this equation is similar to that of convective heat transfer flux $\alpha_{convection}(\theta_{surface} - \theta_{air})$ and is compatible with a system state equation expressing the time evolution of a linear system. Thus, Eq. (2.18) can readily accommodate radiant heat transfer, as demonstrated in the following examples.

Consider the rectangular room outlined in Fig. 2.18. The outside left wall encloses a glazing surface, and each surface is labeled 1–7. Without loss of generality, we assume that the room is air-conditioned and the temperature is retained at 26 °C. If the heat system enclosed by the seven surfaces is space discretized using the CVM or the finite difference method, its system state equation can be expressed by Eq. (2.18), as previously explained. The heat conductance matrix in this case includes the linear radiant heat transfer derived in this section, the convective heat transfer between the wall surfaces and the room, and the heat conduction within the surfaces (i.e., the heat conduction between neighbouring nodes in a wall). Delineating conductive/convective and radiant heat transfer by shading and points, respectively, the matrix $\mathbf{C}$ is dissected as follows:

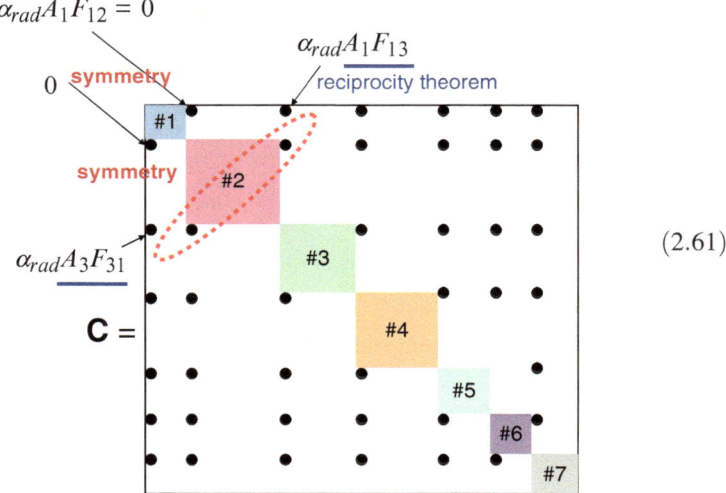

$$\tag{2.61}$$

The elements representing the temperature nodes at each inside-room facing surface are the products of the radiant heat transfer rate, area, and form factor. For example, as shown in (2.61), $\alpha_{rad}A_1 F_{13}$ is incorporated in the internal room surface nodes (i, j) between surfaces 1 and 3, and its transpose $\alpha_{rad}A_3 F_{31}$ appears in the element (j, i). Furthermore, assuming that the reciprocity theorem of the form factor holds, radiant heat transfer can be included in the heat conductance matrix without destroying its symmetry (alternatively, if the heat conductance matrix is

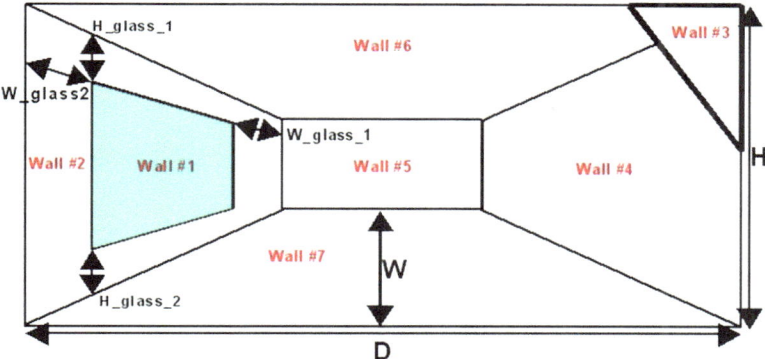

Fig. 2.18 The seven surfaces of a rectangular room

assumed to be symmetric, the reciprocity theorem of the form factor holds). In addition, because the glasing surface 1 and the outside wall of surface 2 reside in the same plane, the form factor between these surfaces is zero. Therefore, their corresponding matrix elements are zero.

2.11 Linear Heat Moisture Transfer Equation

We expect that an analogy exists between heat transfer and humidity propagation through a solid. The latter is driven by water vapour density; thus, if the absolute humidity [g/g'][7] is taken as the potential, a governing equation analogous to the unsteady-state heat transfer Eq. (2.1) is expected. However, steam (gas; vapour) and water (liquid) coexist at normal temperatures, so their transfers require separate treatments (Fig. 2.19). For example, if the temperature within a material increases, some of the water evaporates from the material and the mass of vapour phase increases (assuming that liquid and gaseous phases exist in a state of local equilibrium; the so-called state of local balance); however, because this process involves the latent heat of vaporization L [J/g], it will impact on the heat balance in the region. In such a situation, moisture and heat propagation are inextricably linked, and we must consider *heat moisture transfer*.

In this book, we assume relatively low moisture density inside the material (Fig. 2.20). Such a state is called *hygroscopic*. In a hygroscopic region, the absorbed water is trapped in the interfaces between the material substance and the opening, and is thought to be not easily dislodged (however, under the local equilibrium

---

[7] Humid air is a mixture of dry air (DA) and humidity (moisture). In physical terms, the most appropriate humidity parameter for moisture concentration is the specific humidity[g/g]or mass ratio of water vapour to humid air. In contrast, the absolute humidity [g/g(DA)](or [g/g']) is the mass ratio of water vapour to dry air. Although inconsistent with the true definition of concentration, specific humidity is a standard thermodynamic function. In this book, we also introduce the hygroscopic equation, whose potential function is absolute humidity.

Fig. 2.19 The relationship between gas phase, liquid phase

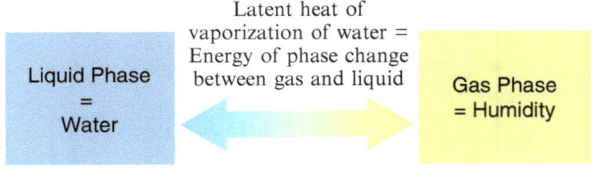

Fig. 2.20 Moisture and humidity inside materials with gaps

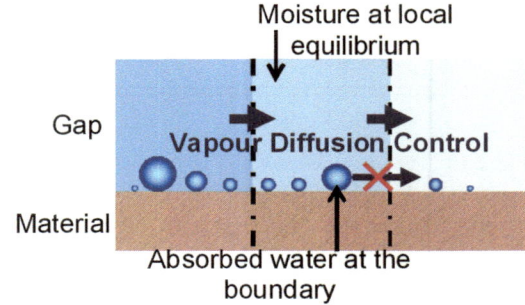

assumption, it can balance the local atmospheric temperature or humidity and can be lost by evaporation or gained by condensation). Hence, diffusive moisture propagation can be considered to occur in the vapour phase only, which radically simplifies the mathematics. However, when considering moisture propagation in the ground or surface condensation, the diffusion of liquid water cannot be ignored; thus, the aforementioned vapour diffusion model is not applicable to situations of high moisture density. However, when dealing with moisture transfer within the walls of a room, it is perfectly appropriate.

Absorbed water w [g] exists in local equilibrium with the temperature θ and the absolute humidity X of the ambient environment. Thus, it can be expressed as

$$w \equiv w(\theta, X). \tag{2.62}$$

The rate of water absorption is then represented by the following total differential:

$$\frac{\partial w}{\partial t} = \frac{\partial w}{\partial \theta} \cdot \frac{\partial \theta}{\partial t} + \frac{\partial w}{\partial X} \cdot \frac{\partial X}{\partial t} \equiv -\nu \cdot \frac{\partial \theta}{\partial t} + \kappa \cdot \frac{\partial X}{\partial t}. \tag{2.63}$$

In the rightmost terms of (2.63), we define $\frac{\partial w}{\partial \theta} \equiv -\nu$ and $\frac{\partial w}{\partial X} \equiv \kappa$. These are physical parameters; the moisture desorption coefficient when a unit volume of the material is exposed to unit temperature change in the local environment [g/(m^3 K)] and the hygroscopic coefficient when a unit volume of the material is exposed to unit absolute humidity change [g/(m^3(g/g'))]. These κ and ν are obtained from the gradient of the water content ratio curve $g(\theta,X)$ of the material at equilibrium.

Fig. 2.21 Representative
water content radio curve
for a material at equilibrium

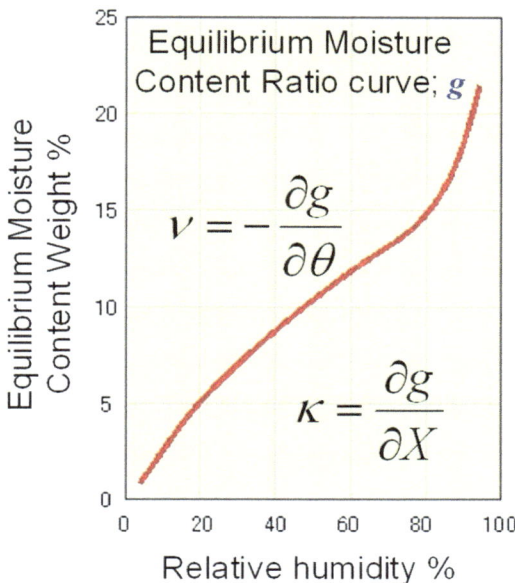

As shown in Fig. 2.21, the equilibrium water content ratio curve plots equilibrium moisture content as a function of relative humidity at constant temperature. These curves can be constructed for representative materials from handbooks listing the physical properties of materials. The experimental apparatus and technique for determining κ and ν is shown in Fig. 2.22. The material is placed on a gravimeter inside a chamber and left as it is at the initial temperature and humidity for the prolonged time period. This "prolonged period" ensures that no further weight change occurs and that the moisture in the material has equilibrated. At this point, the atmospheric absolute humidity is incrementally increased and the weight change is measured. The experiment is terminated once the weight increase has reached sufficient equilibrium. κ is then obtained by dividing the incrementally increased weight (volume) change by the incremental increases in absolute humidity and experimental volume. Similarly, ν is obtained from the weight loss induced by the stepwise temperature increase.

Now, we derive the one-dimensional unsteady-state vapour diffusion heat moisture transfer equation, which is equivalent to the one-dimensional unsteady-state heat transfer Eq. (2.1). As already discussed, the propagation of water vapour should be described by a diffusion equation governed by absolute humidity. The vapour dynamics are then expressed as

$$C' \rho_{air} \frac{\partial X}{\partial t} = \lambda' \frac{\partial^2 X}{\partial x^2}.$$

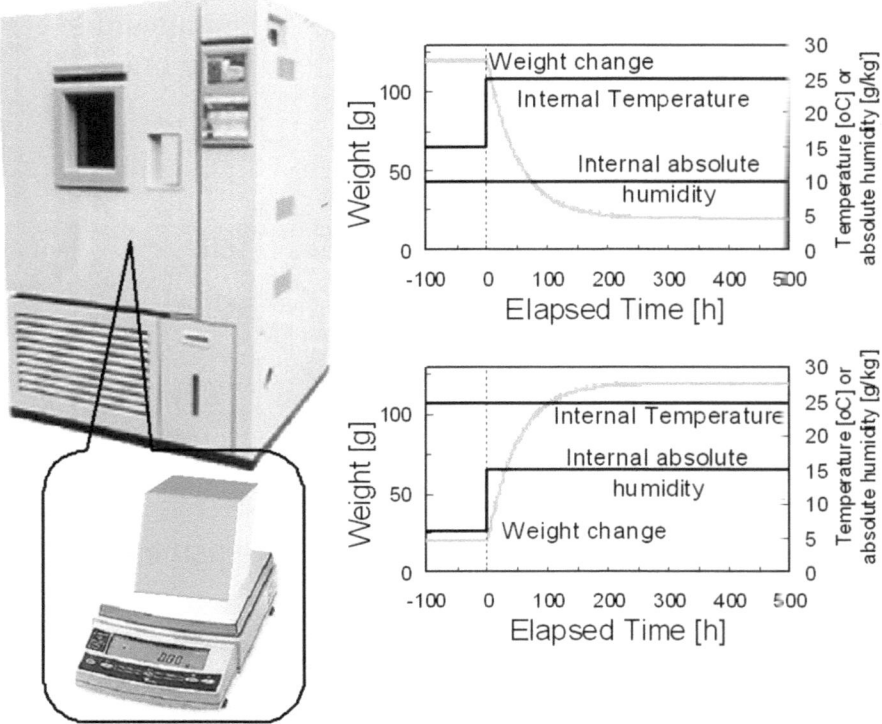

Fig. 2.22 Experimental method for determining κ and ν

where C' is porosity [m³/m³], ρ_{air} is the density of humid air [kg/m³], and λ' is moisture conductivity [g/(ms(g/g'))]. These quantities are analogous to the specific heat, density, and thermal conductivity, respectively, in the heat conduction equation. However, the above expression is incomplete because as explained earlier, the moisture transfer and heat conduction have not been coupled through the latent heat of vaporization. The effect of latent heat is shown schematically in Fig. 2.23. What happens when the weight of absorbed water in the left panel of that figure (which exists in local equilibrium) reduces in an infinitesimal time to that in the right panel? Because water has been lost to evaporate, the concentration of water vapour in the atmosphere, i.e., the absolute humidity, increases. Simultaneously, latent heat of vaporization is removed from the surroundings and the temperature falls. If these physical processes are incorporated into the diffusion equations of heat and moisture conduction, we obtain

$$C_p \rho \frac{\partial \theta}{\partial t} = \lambda \frac{\partial^2 \theta}{\partial x^2} + L \frac{\partial w}{\partial t} \tag{2.64.1}$$

Fig. 2.23 Coupling of heat and moisture

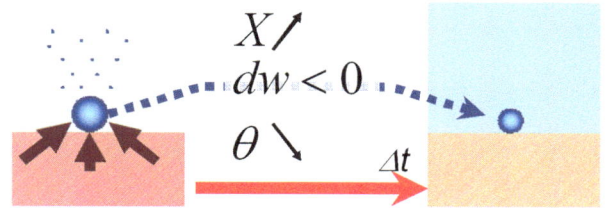

$$C'\rho_{air}\frac{\partial X}{\partial t} = \lambda'\frac{\partial^2 X}{\partial x^2} - \frac{\partial w}{\partial t}. \tag{2.64.2}$$

Note the sign of the second term on the right-hand side of each equation. In the heat transfer equation (2.64.1), water is absorbed from the release of latent heat. In the moisture transfer Eq. (2.64.2), water absorption causes reduction in vapour concentration (absolute humidity). By substituting Eq. (2.63) into Eqs. (2.64.1) and (2.64.2), the hygroscopic one-dimensional heat moisture simultaneous transfer Eqs. (2.65.1) and (2.65.2) are obtained as

$$\left(C_p\rho + L\nu\right)\frac{\partial \theta}{\partial t} = \lambda\frac{\partial^2 \theta}{\partial x^2} + L\kappa\frac{\partial X}{\partial t} \tag{2.65.1}$$

$$\left(C'\rho_{air} + \kappa\right)\frac{\partial X}{\partial t} = \lambda'\frac{\partial^2 X}{\partial x^2} + \nu\frac{\partial \theta}{\partial t}. \tag{2.65.2}$$

Please take another careful look at Fig. 2.21. κ and ν in the above equation are the differentials (with respect to temperature) of absolute humidity and equilibrium moisture content, respectively. In the medium humidity regions of Fig. 2.21, the relationship between equilibrium moisture and relative humidity is reasonably linear, and so κ and ν may be approximated as constants at moderate humidity. If such an approximation is valid in reality, the simultaneous Eqs. (2.65.1) and (2.65.2) become linear, which vastly simplifies the computations. We can now establish the system state equations.

To this end, (2.18) is restated in a slightly different format:

$$\mathbf{M}\frac{d\mathbf{x}}{dt} = \mathbf{Cx} + \mathbf{C_o x_o} + \mathbf{f}. \tag{2.66}$$

Here $\mathbf{x}$ is an unknown variable vector containing the nodes of temperature and absolute humidity, while $\mathbf{x_o}$ is a vector of temperature and absolute humidity at the stipulated nodes. The matrix $\mathbf{C_o}$ holds the heat and moisture flux boundary conditions at the stipulated nodes. The matrices $\mathbf{M}$ and $\mathbf{C}$ are called the *expansion capacitance matrix* and the *expansion conductance matrix*, respectively. The elements of these vectors and matrices are explained in the following example.

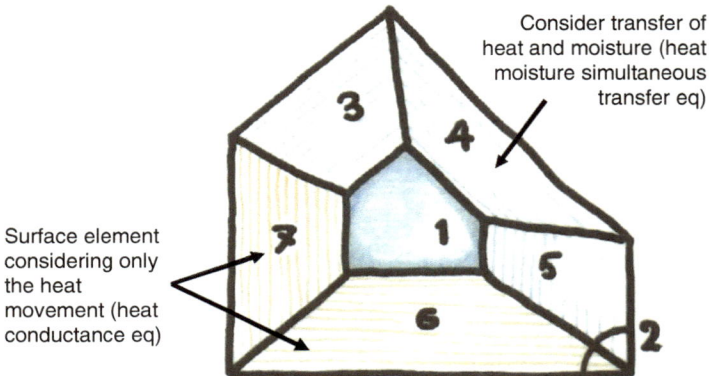

Fig. 2.24 Room model solved by hygrothermal equations

Consider a single room, as shown in Fig. 2.24. Suppose that the temperature and humidity of the room are restrained at the stipulated nodes. In other words, the heat conductance and moisture transfer equations at each surface node are given by (2.65.1) and (2.65.2). The surfaces are numbered as shown in the figure, but those at which heat conductance alone is relevant are numbered as a support. For example, heat transfer through glass and metal surfaces is more appropriately modelled by Eq. (2.1) than by (2.65.1) and (2.65.2).

When space is discretized by the control volume element method in the finite difference formulation, the vectors and matrices of Eq. (2.66) contain the following elements. In particular, note the elements of **M**.

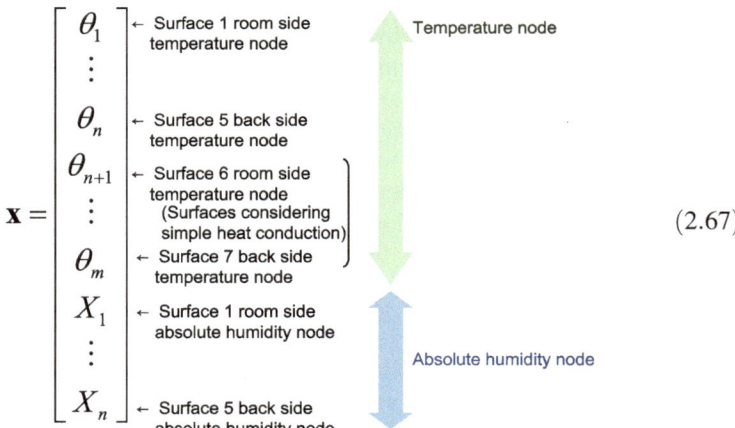

$$(2.67)$$

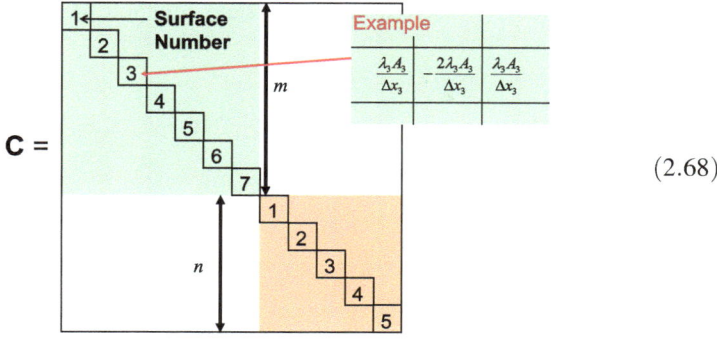

$$\text{(2.68)}$$

$$\mathbf{C_o} = \qquad\qquad \text{(2.69)}$$

$$\mathbf{X_o} = \begin{array}{c} \theta_{out} \\ \theta_{neighbor1} \\ \vdots \\ X_{out} \\ X_{neighbor1} \\ \vdots \end{array} \qquad \text{(2.70)}$$

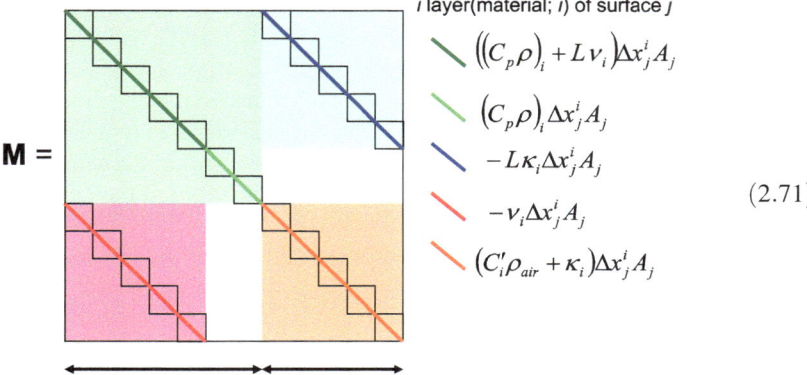

$$\text{(2.71)}$$

Hence, the spatially discretized system state equation has the same format as Eq. (2.29), and

$$
\mathbf{x}^{i+1} = \left[\frac{1}{\Delta t}\mathbf{M} - k\mathbf{C} \right]^{-1}
$$
$$
\left\{ \left[\frac{1}{\Delta t}\mathbf{M} + (1-k)\mathbf{C} \right]\mathbf{x}^i + \mathbf{C_o}\left\{ (1-k)\mathbf{x}_\mathbf{o}^i + k\mathbf{x}_\mathbf{o}^{i+1} \right\} + \left\{ (1-k)\mathbf{f}^i + k\mathbf{f}^{i+1} \right\} \right\},
$$

$$(2.72)$$

where k is an arbitrary real number in $k \in [0,1]$, which takes values 0, 1/2, and 1 for forward difference, Crank–Nicolson, and backward difference schemes, respectively.

2.12   Calculations of Heat Load and Natural Room Temperature

Here we apply the calculations of heat load and natural room temperature to the system state equations. In fact, the latter has been already described, but for the reader's benefit, both are described to emphasize the difference between them.

In Sect. 2.1, heat conduction was shown to be analogous to the equation of motion in particle dynamics. Take a look at Fig. 2.25, demonstrating a ball flicked on a desk. The calculation of natural room temperature is equivalent to the situation in the upper panel of the figure. In other words, we seek the change in the room temperature in response to various heat inputs. The unknown quantities are the room temperature and the initial speed of the flicked ball. (Although the dimensions are different, the distance over which the ball rolls by inertia is relevant to our analogy.) Conversely, for the load calculation, the room temperature is known and is retained constant at 26 °C (or 28 °C if cooling). In this case, we solve for the thermal requirement to meet the set room air temperature; namely heat extraction (cooling load) or heat supply (heat load). The equivalent mechanical problem in the lower panel calculates the force required to completely halt the ball when the ball is flicked with the same force f as in the upper panel.

More specifically, we compute the changes of natural room temperature and heat load within a single room, as in Fig. 2.26. The room is identical to that of Fig. 2.18, but here we consider that fresh external air is introduced at a circulation rate of n [1/s] and that h watts of heat are generated within the room. If the circulation rate is multiplied by the room volume, the circulation volume (ventilation rate) [m³/s] is obtained. Both circulation volume and heat generation are suitable parameters for determining the heat load. The former is called the external air load due to circulation (including draughts) and the latter is the internal generated heat load introduced by human body heat or interior electrical equipment.

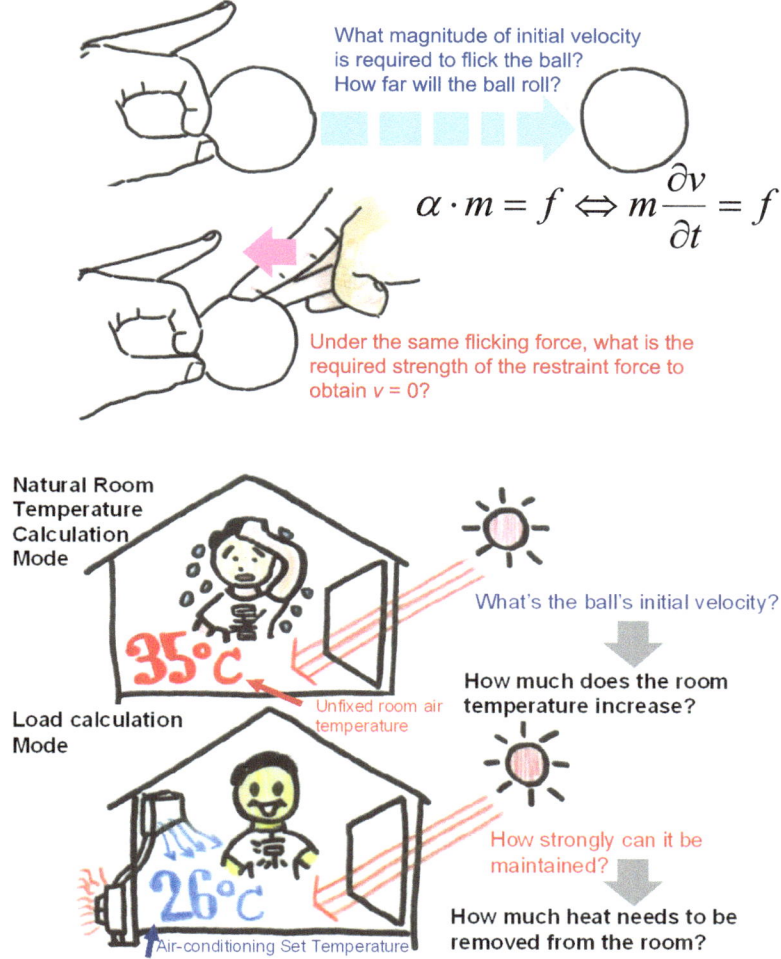

Fig. 2.25 Analogy to classical mechanics: calculation of heat load and natural room temperature

The heat balance equation for the evolution of natural room temperature at room temperature node θ_r becomes

$$\underbrace{V_r(C_p\rho)_{air}\frac{\partial \theta_r}{\partial t}}_{\frac{[m^3][Jm^{-3}K^{-1}][K]}{[s]} = [W]} = \underbrace{\sum_{i \in \{wall\}} A_i \alpha_{conv}^i (\theta_{surface}^i - \theta_r)}_{[m^2][Wm^{-2}K^{-1}][K] = [W]} + \underbrace{\frac{nV_r(C_p\rho)_{air}(\theta_{out} - \theta_r)}{[s^{-1}][m^3][Jm^{-3}K^{-1}][K] = [W]}}_{} + \underbrace{\frac{h}{[W]}}_{}$$

$$(2.73.1)$$

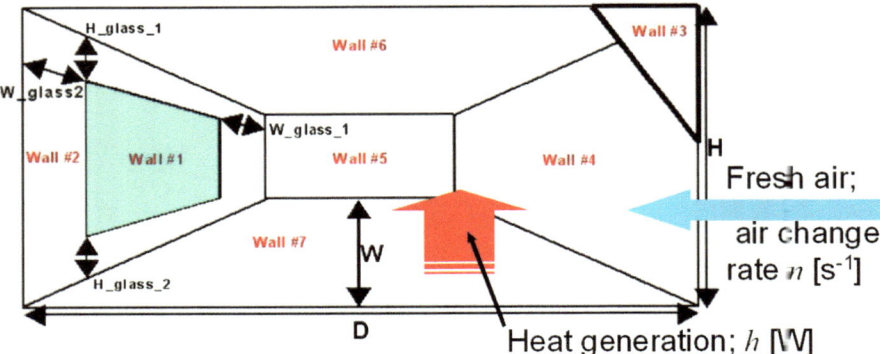

Fig. 2.26 Temperature and heat-load calculations applied to a natural room

The reader should confirm that the dimension of all underlined terms in (2.73.1) is [W]. The first, second, and third terms on the right-hand side describe the heat gained by convective heat transfer between wall surfaces, the heat gained by ventilation, and the internally generated heat, respectively.

In contrast, in the heat load formulation of the heat balance equation, the room temperature is retained at θ_{set}, and the cooling load term H_{ex} [W] becomes an unknown variable in the following expression:

$$V_r \left(C_p\rho\right)_{air} \frac{\partial \theta_r}{\partial t} = \sum_{i \in \{wall\}} A_i \alpha_{conv}^i \left(\theta_{surface}^i - \theta_{set}\right)$$

$$+ n V_r \left(C_p\rho\right)_{air} \left(\theta_{out} - \theta_{set}\right) + h - H_{ex}. \qquad (2.73.2)$$

At the start of cooling, (2.73.2) reduces to

$$V_r \left(C_p\rho\right)_{air} \frac{\partial \theta_r}{\partial t} = V_r \left(C_p\rho\right)_{air} \frac{\theta_r^{j-1} - \theta_{set}}{\Delta t}. \qquad (2.73.3)$$

Under continuous operation, (2.73.3) further reduces to

$$V_r \left(C_p\rho\right)_{air} \frac{\partial \theta_r}{\partial t} = 0. \qquad (2.73.4)$$

If the heat capacitance of air is assumed sufficiently small relative to the wall heat capacity, it can be ignored, and the left-hand side of Eq. (2.73.2) permanently vanishes.[8] In this case, the system state equation is that of Eq. (2.18) regardless of whether the room temperature is calculated in a natural or air-conditioned environmen. The vector and matrix elements in each case are elucidated below.

---

[8] The system state equation may also be computed by incorporating Eq. (2.73.2), without making this approximation (more specifically, $C_o \theta_o$ may be incorporated). This assumption was introduced to simplify the explanation.

First we consider the calculation of natural room temperature.

$$\theta = \quad\quad\quad\quad\quad\quad\quad\quad (2.74)$$

Surface 1 room side surface temperature node

N_{total}

Surface 7 back side surface temperature node

θ_r

The N_{total} th row of the unknown vector holds the room temperature θ_r.

$$M = \quad\quad\quad\quad\quad\quad\quad\quad (2.75)$$

$V_r(C_p\rho)_{air}$

The heat capacities $V_r(C_p\rho)_{air}$ are entered into element (N_{total}, N_{total}) of the heat capacitance matrix.

$$f = \quad\quad\quad\quad\quad\quad\quad\quad (2.76)$$

h

The heat input boundary condition vector f contains appropriate values. The heat generated in the room is held in row N_{total}. The coloured dots* indicate heat sources such as solar transmission through window surfaces distributed across the surface temperature nodes*.

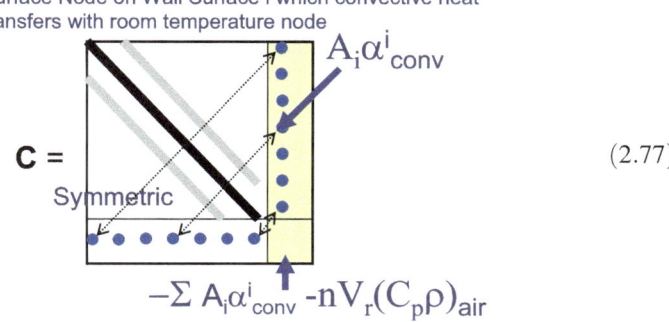

Surface Node on Wall Surface i which convective heat transfers with room temperature node

$A_i\alpha^i_{conv}$

$$C = \quad\quad\quad\quad\quad\quad\quad\quad (2.77)$$

Symmetric

$$-\Sigma\, A_i\alpha^i_{conv} - nV_r(C_p\rho)_{air}$$

The wall surface components of the heat conductance matrix contain the heat transfer conductance between neighbouring nodes (thick black diagonal line and gray diagonal lines in (2.77)). The convective heat transfers $A_i \alpha_{conv}^i$ (blue points) of each room-facing wall surface occupy the column N_{total} (where A_i is the surface area of the i th surface and α_{conv}^i is the convection heat transfer rate of surface i). These quantities also appear in the row N_{total}, rendering (2.77) a symmetric matrix.

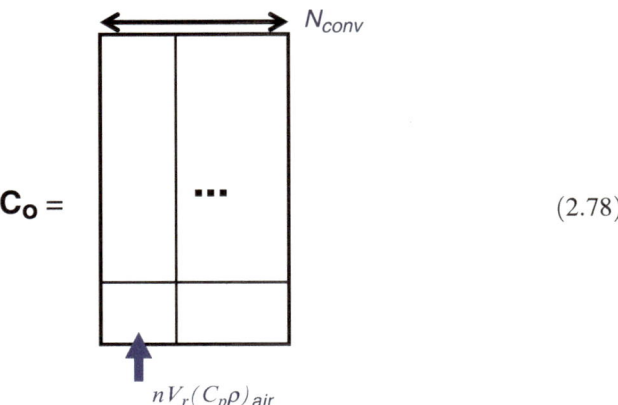

$$\mathbf{C_o} = \qquad\qquad (2.78)$$

The number of columns in $\mathbf{C_o}$ is the number of stipulated temperature nodes N_{conv}. The first column expresses the heat conductance of the external air*. Therefore, row N_{total} holds the conductance of ventilation* $nV_r(C_p\rho)_{air}$.

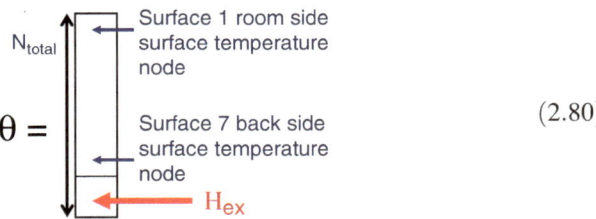

$$\boldsymbol{\theta_o} = \qquad\qquad (2.79)$$

The number of rows in $\boldsymbol{\theta_o}$ is N_{conv}. The first row holds the external temperature θ_{out}.

Next, we construct the matrix elements for the heat load calculation.

$$\theta = \qquad\qquad (2.80)$$

N_{total}

Surface 1 room side surface temperature node

Surface 7 back side surface temperature node

H_{ex}

In this case, the N_{total} th row of the unknown vector holds the heat load H_{ex}.

$$\mathbf{M} = \qquad\qquad\qquad (2.81)$$

Because Eq. (2.73.4) is assumed, the element (N_{total}, N_{total}) in the heat capacitance matrix is zero. The heat input boundary condition vector $\mathbf{f}$ is that of Eq. (2.76).

$$\mathbf{C} = \qquad\qquad\qquad (2.82)$$

Surface node on wall i which has
convective heat transfer with room $A_i \alpha^i_{conv}$

The N_{total} th column of the heat conductance matrix constructed for natural room temperature is now shifted to the $(N_{conv} + 1)$ th column of $\mathbf{C_o}$ and the symmetry is broken. Then -1 is substituted into element (N_{total}, N_{total}) of $\mathbf{C}$ so that the unknown heat load H_{ex} can appear on the left side of Eq. (2.18).

$$\mathbf{C_o} = \qquad\qquad\qquad (2.83)$$

$$nV_r(C_p\rho)_{air} - \Sigma\, A_i\alpha^i_{conv} - nV_r(C_p\rho)_{air}$$

$\mathbf{C_o}$'s $(N_{total}, N_{conv} + 1)$ element, similarly to the usual $\mathbf{C}$'s diagonal element, values multiplying the row sum of $\mathbf{C_o}$ and $\mathbf{C}$ by -1 is entered.

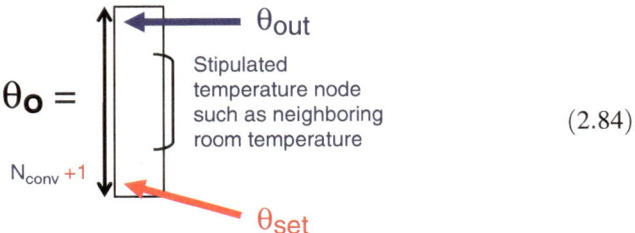

$$\boldsymbol{\theta_o} = \quad \begin{array}{c} \theta_{out} \\ \text{Stipulated temperature node such as neighboring room temperature} \\ N_{conv} +1 \quad \theta_{set} \end{array} \qquad (2.84)$$

$\boldsymbol{\theta_o}$ contains $N_{conv} + 1$ rows (recall that in the natural room temperature setup, $\boldsymbol{\theta_o}$ contains N_{conv} rows). Row $N_{conv} + 1$ holds θ_{set}.

From the above analysis, we should understand that whether calculating the natural room temperature or the heat load, by manipulating the contents of the component matrices and vectors, we can obtain a unified discretization equation in the form of Eq. (2.18).

2.13   Numerical Simulation of a Single Room Model

In this section, to consolidate the items discussed so far, the system state equations, formulated in terms of the natural room temperature and heat load, will be solved numerically in Fortran.

The system is a single room as shown in Fig. 2.27. The room is intermittently air-conditioned from 9 a.m. to 5 p.m. We seek the changes in the cooling load during this period [W/m^2] and the natural room temperature [°C] outside this time. The south face of the room is external facing and contains a glazing window made of 3 mm stratum glass. The dimensions of each section are shown in the figure and are assigned variable names such as *HEI* and *WID* [m] in the program. Appropriate parameter values should be selected by the reader. Above and below stairs, the north, east, and west sides touch the neighbouring room. External air is introduced at a circulation rate of *RNV* [1/s].

The external temperature θ_o and heat flux of solar radiation are given in Table 2.1. Data in this table are weather data for air-conditioning design under presumed conditions of a harsh summer.[9] Solar radiation [kcal/(m^{2}h)] is assumed incidental to the south surface. Note the change from engineering units to the SI unit system (only the relationship 1 [W] = 0.86 [kcal/h] should be remembered). The blank cells in the solar radiation data imply that solar irradiation was zero or not obtained.

---

[9] These weather data consist of hourly temperature and hourly solar radiation applied with excess frequency ratio 2.5 % (so-called TAC 2.5 [Technical Advisory Committee of ASHRAE]; meaning top 2.5 % highest temperature and radiation rate in last 10 years as the statistical samples), which, if used in cooling design, will overestimate the device capacity of refrigerators and air-conditioning units. This occurs because the time-series constructed from the weather data distorts the real-life events.

Fig. 2.27 Room model of programming example

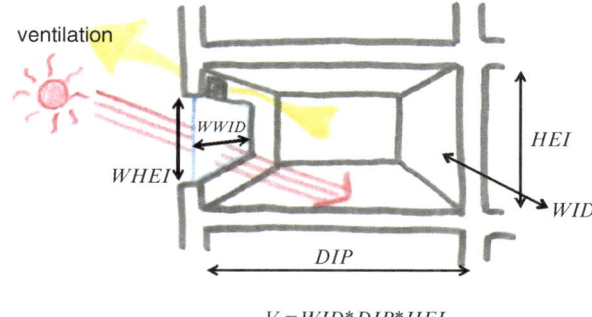

$$V_r = WID * DIP * HEI$$

Now the seven-surface room (including the glass window surface) is radically deformed into that of Fig. 2.28. This snow-hut structure, called "Kamakura" in Japan, is similar to the structure introduced in Sect. 2.9. The external walls and open glass surface are modelled intact, while the ceiling, floor, and neighbouring dividing walls are modelled as a single internal wall (preserving the total surface area). The external wall is assumed 3-layered (from the external side, bricks, insulator, concrete), while the inside walls are assumed single-layered (concrete). The heat capacity of the glass is considerably smaller than that of the other wall elements and is hence ignored; that is, like the surface temperature nodes, the glass surface nodes are assigned no heat capacity, ○. Unknown values are labelled with the following node numbers: ① external room side surface, ⑦ external surface, ⑧ glass surface, ⑨ internal room-facing surface, ⑬ neighbouring side surface, ⑭ room temperature nodes. Furthermore, while air-conditioning is operating, the unknown value in ⑭ is the cooling load. The reader is encouraged to research (using appropriate resources) the values of inside and outside heat transfer coefficient α_o, α_r [W/(m² K)], solar transmissibility of glass TAU_g, absorptivity ABS_g, solar absorptivity of the external wall ABS_w, and solar absorptivity of the internal walls TAU_IW. In addition, the reader should appreciate the thermophysical properties of the material and gauge the values of other required constants. The solar radiation penetrating the open glass surface is assumed to be fully incident on the internal wall surface ($ABS\_IW = 1$).

Let us now construct the unknown variable vector and the stipulated temperature node vector of the system state Eq. (2.18). First, in the natural room temperature mode, the 14th unknown value is the room temperature θ_{14}. The stipulated temperature node vector contains the external temperature and the neighbouring room temperature θ_o. The temperature of the room at time $t - 1$ becomes the neighbouring room temperature θ_b at time t^*.

$$\theta = \boxed{} \quad \leftarrow \theta_{14} \tag{2.85}$$

Table 2.1 Summer air-conditioning weather data; external temperature with excess significant level 2.5 % [°C]; south-facing vertical solar irradiation [kcal/(m² h)]

	1	2	3	4	5	6	7	8	9	10	11	12	13	14	15	16	17	18	19	20	21	22	23	24
θ_o	27.6	27.4	27.2	26.9	26.8	27.0	28.1	29.4	30.7	31.7	32.5	33.1	33.4	33.4	33.1	32.4	31.6	30.7	30.0	29.3	28.8	28.4	28.1	27.9
I_{sol}					8	26	35	54	137	201	240	248	227	176	103	38	32	21						

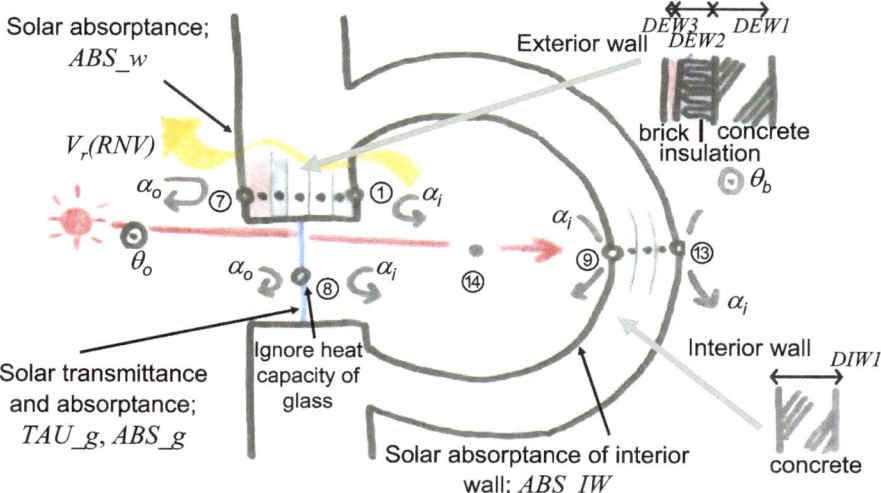

Fig. 2.28 Deformed single room model

$$\theta_\mathbf{o} = \begin{array}{|c|} \hline \leftarrow \theta_o \\ \hline \leftarrow \theta_b \\ \hline \end{array} \qquad (2.86)$$
$$\mathbf{||}$$
$$\theta_{14|\text{previous time-step}}$$

In the load calculation mode, the 14th unknown value is the heat load H_{ex}. The 3-row stipulated temperature node vector contains the external temperature θ_o, the neighbouring temperature θ_b, and the air-conditioning set temperature θ_{set}.

$$\theta = \begin{array}{|c|} \hline \\ \\ \hline \leftarrow Hex \\ \hline \end{array} \qquad (2.87)$$

$$\theta_\mathbf{o} = \begin{array}{|c|} \hline \leftarrow \theta_o \\ \hline \leftarrow \theta_b = \theta_{14|\text{previous time-step}} \\ \hline \leftarrow \theta_{set} \\ \hline \end{array} \qquad (2.88)$$

As previously explained, the weather data are collected over one day. In the simulation, these data are applied repeatedly as the boundary conditions, and

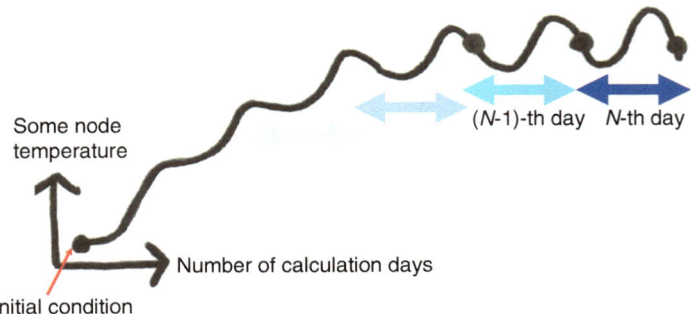

Some node
temperature

(N-1)-th day   N-th day

Number of calculation days

Initial condition

Fig. 2.29 Daily steady-state calculation

calculations are repeated until the effect of the initial conditions diminishes and a
daily steady-state solution is obtained (see Fig. 2.29).

The Fortan source code is provided in the following pages. A subroutine is
merged immediately after the main section; then, if the code is collectively com-
piled, an executable file is created. The program outputs the 24-h series of the above
daily steady-state solution.

```fortran
      program AC_Load_Temp
      parameter(nn=14, mm=2) ! nn; Total number of nodes/
      mm; Number of stipulation nodes in natural room
c  Temperature calculation mode (-> m+1 in heat load
   calculation mode)
c    【Array declaration】
c  Natural room temperature calculation mode [M],
   [C],  [Co]
      real*4 M_Temp(nn,nn),C_Temp(nn,nn),Co_Temp(nn,mm),
c        [A],   [B],   [A^-1]
     *  A_Temp(nn,nn),B_Temp(nn,nn),Ainv_Temp(nn,nn),
c        {θo}
     *  theta_o_Temp(mm)
c  Heat load calculation mode [M],    [C],       [Co]
      real*4 M_Load(nn,nn),C_Load(nn,nn),Co_Load(nn,mm+1),
c        [A],   [B],   [A^-1]
     *  A_Load(nn,nn),B_Load(nn,nn),Ainv_Load(nn,nn),
c        {θo}
     *  theta_o_Load(mm+1)
c  Vector of unknown variables θ,Vector of boundary
   condition given by heat flux {f}
```

```
        dimension theta(nn),f(nn),
c  Vector of unknown variables for daily steady-state
   calculation; saving 24 hors data in the previous day
   *   theta_24h(nn)
c  External air temperature, south-facing vertical
   solar radiation
        dimension air(24),solar(24)
c  Vector for interim output in calculation process
   {x1},{x2},{x3}
        dimension x1(nn),x2(nn),x3(nn)
c   【Definition of output files】
        open(10,file='result.csv')
c   【Definition of assumed data】
c  Thermo-physical properties
c           In this program, all data is given in
            Engineering Unit.
c           Thus, all variables should be transferred to
            SI unit when outputting.
c  λ : thermal conductivity/ brick, insulation,
   concrete, glass [kcal/(mh°C)]
        data RAMM, RAMF, RAMC, RAMgla/ 0.55, 0.032, 1.2, 0.67/
c  Cpρ : volumetric specific heat/ brick, insulation,
   concrete, glass [kcal/(m^3°C)]
        data GAMM, GAMF, GAMC/ 332., 8.4, 462./
c  volumetric specific heat of humid air [kcal/(m^3°C)]
        GAMA=1.205*0.24
        ALPI = 10. ! convective heat transfer coefficient at
        interior surface [kcal/(m^2h°C)]
        ALPO = 20. ! convective heat transfer coefficient at
        exterior surface [kcal/(m^2h°C)]
c  dimension of the room/ frontage, depth, height, window
   width, window height [m]
        data DIP, WID, HEI, WWID, WHEI/ 3.6, 2.13, 2.6, 3.0,1.0 /
c  wall thickness/ layer #3 of exterior wall(brick),
   #2(insulation), #2(concrete), interior
   wall(concrete), window(glass) [m]
        data DEW3, DEW2, DEW1, DIW1, Dgla/0.010, 0.10, 0.30,
        0.150, 0.003/
c  exterior air temperature (24-hours variation)
        data air /27.6, 27.4, 27.2, 26.9, 26.8, 27.0, 28.1,
        29.4, 30.7,
```

```
      &               31.7, 32.5, 33.1, 33.4, 33.4, 33.1, 32.4,
                      31.6, 30.7,
      &               30.0, 29.3, 28.8, 28.4, 28.1, 27.9/
c  south-facing vertical solar radiation (24-hours
   variation) [kcal/m^2]
   data solar/0.0, 0.0, 0.0, 0.0, 8.0, 26.0, 35.0, 54.0,
   137.0, 201.,
      &               240.0, 248.0, 227.0, 176.0, 103.0, 38.0,
                      32.0, 21.0,
      &               0.0, 0.0, 0.0, 0.0, 0.0, 0.0/
c  air change rate [1/h]
   RNV = 0.3
c  solar absorptance of exterior wall, transmittance of
   glass, absorptance of glass, absorptance of interior
   wall [ND]
   data ABS_w, TAU_g, ABS_g, ABS_IW/ 0.8, 0.7, 0.1, 1.0/
c  time discretization step Δt[h], threshold to evaluate
   daily steady-state ε[°C], cooling set-point
   temperature[°C]
   data delt, Eps, theta_set/ 1. , 0.01, 28./
c  space discretization step Δx [m]
   dxE3 = DEW3/1.   ! layer #3 of external wall (brick) [m]
   dxE2 = DEW2/1.   ! layer #2 (insulation)
   dxE1 = DEW1/3.   ! layer #1 (concrete)
   dxI1 = DIW1/3.   ! interior wall (concrete)
c  other assumptions
   Vr = DIP*WID*HEI          ! volume of the room[m^3]
   AP = 2*(DIP*WID+DIP*HEI)+WID*HEI
                             ! area of interior wall [m^2]
   AO = WID*HEI-WHEI*WWID ! area of exterior wall [m^2]
   AG = WHEI*WWID            ! area of glazing window [m^2]
   AF = DIP*WID              ! floor area [m^2]
c  defnition of starting and terminating cooling
   operation
   j_on=9     ! on; starting cooling operation
   j_off=17   ! off; terminating cooling operation
c  【Space discretization】
c  Heat capacitance matrix [M]
c  《For natural room temperature calculation mode》
   call CLEAN(nn,nn,M_Temp) ! initializing M_Temp
   do i=2,4
            M_Temp(i,i)=GAMC*dxE1*AO
                  ! layer #1 of exterior wall (concrete)
```

```
      enddo
      M_Temp(5,5)=GAMF*dxE2*AO ! layer #2 of exterior wall
                                      (insulation)
      M_Temp(6,6)=GAMM*dxE3*AO  ! layer #3 of exterior wall
                                      (brick)
      do i=10,12
             M_Temp(i,i)=GAMC*dxI1*AP
                                    ! interior wall (concrete)
      enddo
      M_Temp(14,14)=GAMA*Vr
                              ! node of room air temperature*
c    《For heat load calculation mode》
      call CLEAN(nn,nn,M_Load) ! Initializing M_Load
      call equal(nn,nn,nn,nn,M_Load,M_Temp)
                              ! Copy M_Temp to M_Load as is
      M_Load(14,14)=0.
c  Matrix Co representing  boundary  condition  between
   heat conductance matrix C and stipulated nodes
c    《For natural room temperature calculation mode》
c  [C]
      call CLEAN(nn,nn,C_Temp)
      C_Temp(1,2)=RAMC/(dxE1/2.)*AO ! exterior wall
      C_Temp(2,3)=RAMC/dxE1*AO
      C_Temp(3,4)=RAMC/dxE1*AO
      C_Temp(4,5)=1./((0.5*dxE1/RAMC)+(0.5*dxE2/RAMF))*AO
                              ! composite conductance*
      C_Temp(5,6)=1./((0.5*dxE2/RAMF)+(0.5*dxE3/RAMM))*AO
                              ! composite conductance*
      C_Temp(6,7)=RAMM/(dxE3/2.)*AO
      C_Temp(9,10)= RAMC/(dxI1/2.)*AP ! internal Wall
      C_Temp(10,11)=RAMC/dxI1*AP
      C_Temp(11,12)=RAMC/dxI1*AP
      C_Temp(12,13)=RAMC/(dxI1/2.)*AP
      C_Temp(1,14)=ALPI*AO ! convective heat transfer
                              between nodes of interior
                              surface of exterior wall and
                              room air
      C_Temp(8,14)=1./(1./ALPI+0.5*Dgla/RAMgla)*AG
                              ! convective heat transfer
                              between nodes of interior
                              surface of window
```

```
c                                 and room air (composite
                                  conductance)
    C_Temp(9,14)=ALPI*AP ! convective heat transfer
                                  between nodes of interior
                                  surface of interior wall and
                                  room air
c                                 Upper triangle->copy to
                                  lower triangle
    do i=1,nn
            do j=1,nn
                    if(i.lt.j)C_Temp(j,i)=C_Temp(i,j)
            enddo
    enddo
c [Co]
    call CLEAN(nn,mm,Co_Temp)
    Co_Temp(7,1) = ALPO*AO ! convective heat transfer
                                  between nodes of external
                                  surface of exterior wall and
                                  external air
    Co_Temp(8,1) = 1./(1./ALPO+0.5*Dgla/RAMgla)*AG
                                  ! convective heat transfer
                                  between nodes of external
                                  surface of
c                                 window and external air
                                  (composite conductance)
    Co_Temp(14,1) = RNV*Vr*GAMA ! Conductance through
                                       ventilation
    Co_Temp(13,2) = ALPI*AP ! convective heat transfer
                                  between nodes of another
                                  surface of interior wall
                                  and
c                                 neighboring room air
c  for diagonal elements of [C]
    do i=1,nn
            C_Temp(i,i)=0.
            do j=1,nn
                if(i.ne.j)C_Temp(i,i)=C_Temp(i,i)
                +C_Temp(i,j)
            enddo
            do j=1,mm
                C_Temp(i,i)=C_Temp(i,i)+Co_Temp(i,j)
            enddo
            C_Temp(i,i)=-C_Temp(i,i)
```

```
     enddo
c    《For heat load calculation mode》
c    [C] & [Co]
     call CLEAN(nn,nn,C_Load)
     call equal(nn,nn,nn,nn,C_Load,C_Temp)
     call CLEAN(nn,mm+1,Co_Load)
     call equal(nn,mm,nn,mm,Co_Load,Co_Temp)
     do i=1,nn-1
     Co_Load(i,mm+1)=C_Temp(i,nn)
                              ! IC(i,14) for natural room
                                temperature calculation mode
                                should be moved to
     C_Load(i,nn)=0.    ! Co(i,3) for heat load
                                calculation mode.
                              ! Thus, C(i,14) for heat load
                                calculation mode must be 0.
c                             See the text around Eq.(2.82).
     enddo
     Co_Load(nn,mm+1)=C_Temp(nn,nn)
                              ! Co(14,3) for heat load calcula-
                                tion mode should be moved to
c                             C(14,14) for natural room
                                temperature calculation mode.
c                             See Eq.(2.83).
     C_Load(nn,nn)=-1. ! C(14,14) for heat load
                                calculation mode is -1.
                                See Eq.(2.82).
c【Time discretization】CAUTION;  This code is based on
                                backward  difference
                                method.
     call CLEAN(nn,nn,A_Temp)
     call CLEAN(nn,nn,B_Temp)
     call CLEAN(nn,nn,Ainv_Temp)
     call CLEAN(nn,nn,A_Load)
     call CLEAN(nn,nn,B_Load)
     call CLEAN(nn,nn,Ainv_Load)
     do i=1,nn
       do j=1,nn
         A_Temp(i,j)=(1/delt)*M_Temp(i,j)-C_Temp(i,j)
         Ainv_Temp(i,j)=A_Temp(i,j)
         B_Temp(i,j)=M_Temp(i,j)/delt
         A_Load(i,j)=(1/delt)*M_Load(i,j)-C_Load(i,j)
```

```
        Ainv_Load(i,j)=A_Load(i,j)
        B_Load(i,j)=M_Load(i,j)/delt
      enddo
    enddo
    call MATINV(Ainv_Temp,nn,nn)
    call MATINV(Ainv_Load,nn,nn)
c【Initial temperature assignment*】
    do i=1,nn
      theta(i)=0.
      theta_24h(i)=theta(i)
    enddo
    theta_room=theta(nn) ! Room temperature
c【Time step-by-step calculation loop 】
    do iday=1,100 ! loop for daily steady-state
    calculation; upper limit -> 100 days
      do j=0,23 ! time step loop; calculating (j+1) time
      step because of backward difference method
        if(j+1.lt.j_on.or.j+1.gt.j_off)then
          ! Natural room temperature calculation mode
          call PROVEM(nn,nn,B_Temp,theta,x1,nn,nn)
c             ↑ {x1}={B}{θ}
          theta_o_Temp(1) = air(j+1)
          theta_o_Temp(2) = theta_room
          call PROVEM(nn,mm,Co_Temp,theta_o_Temp,
          x2,nn,mm)
c             ↑{x2}={Co}{θo}
          call CLEANV(nn,f) ! initializing {f} and fix
          initial {f}
          f(7)=ABS_w*solar(j+1)*AO
            ! Solar radiation absorbed by external wall
          f(8)=ABS_g*solar(j+1)*AG
            ! Solar radiation absorbed by glass
          f(9)=TAU_g*ABS_IW*solar(j+1)*AG
            ! Transmitted radiation through glazing
            window
c           is absorbed by interior wall
          do i=1,nn ! {x3}={x1}+{x2}+{f}
            x3(i)=x1(i)+x2(i)+f(i)
          enddo
          call PROVEM(nn,nn,Ainv_Temp,x3,theta,nn,nn)
            ! {θ}={A^1}{x3}
          theta_room=theta(nn) ! Room temperature
```

```
            HEX=0.                          ! Heat load
            if(j+1.eq.j_on-1)theta(nn)=HEX
               ! Specific disposition for variable switching
C                  when cooling operation turns on.
          else ! Heat load calculation mode
            call PROVEM(nn,nn,B_Load,theta,x1,nn,nn)
               ! {x1}={B}{θ}
            theta_o_Load(1) = air(j+1)
            theta_o_Load(2) = theta_room
            theta_o_Load(3) = theta_set
            call PROVEM(nn,mm+1,Co_Load,theta_o_Load,
            x2,nn,mm+1)
C                  ↑ {x2}={Co}{θo}
            call CLEANV(nn,f) ! initializing {f}, and
            presuming each of elements of {f}
            f(7)=ABS_w*solar(j+1)*AO
               ! absorbed solar radiation at exterior wall
            f(8)=ABS_g*solar(j+1)*AG
               ! absorbed solar radiation at glazing window
            f(9)=TAU_g*ABS_IW*solar(j+1)*AG
               ! transmitted solar radiation through
                 window is absorbed
C                  at interior wall
            do i=1,nn ! {x3}={x1}+{x2}+{f}
               x3(i)=x1(i)+x2(i)+f(i)
            enddo
            call PROVEM(nn,nn,Ainv_Load,x3,theta,nn,nn)
               ! {θ}={A^1}{x3}
            theta_room=theta_set ! room temperature
            HEX=theta(nn)/0.86/AF
               ! heat load per floor area, expressed with SI
                 unit [W/m^2]
            if(j+1.eq.j_off)theta(nn)=theta_room
               ! Specific disposition for variable
                 switching
C                  when cooling operation turns off.
          endif
          write(10,100)iday,j+1,(theta(i),i=1,nn-1),
          theta_room,HEX ! output at each time step
        enddo
C     evaluation whether it attains to daily steady state
      or not
```

```
      do i=1,nn-1
         if(abs(theta_24h(i)-theta(i)).gt.Eps)goto 51
      enddo
      goto 52
 51   continue
      do i=1,nn-1
         theta_24h(i)=theta(i)
      enddo
    enddo
c  This is the end of Time step-by-step calculation loop.
 52   continue
      close(10)
100  format(2(i2,','),100(f9.3,','))
101  format(100(f9.3,','))
      stop
      end

c  Hereinafter, subroutines
c**************************************************
      subroutine CLEAN(M,N,W)
c     Initializing matrix W(M,N)
      DIMENSION W (M,N)
      DO 10 I=1,M
         DO 11 J=1,N
            W(I,J)=0.0
 11      CONTINUE
 10   CONTINUE
      RETURN
      END
c**************************************************
      SUBROUTINE MATINV(AI,NN,NNN2)
c     Calculating inverse matrix of AI(nn,nn), and its
      result is overwritten in same AI
c     CAUTION; array declaration is AI(NN2,NN2),
               irrespective to the size of your project; NN
      DIMENSION AI(NNN2,NNN2),IND(1000)
      DO 102 K=1,NN
 102 IND(K)=K
      DO 103 K=1,NN
      W=0.
      DO 104 I=K,NN
      IF(ABS(AI(I,1)).LE.W) GO TO 104
```

```
      W=ABS(AI(I,1))
      IR=I
104 CONTINUE
      IF(IR.EQ.K) GO TO 106
      DO 107 J=1,NN
      W=AI(K,J)
      AI(K,J)=AI(IR,J)
107 AI(IR,J)=W
      M=IND(K)
      IND(K)=IND(IR)
      IND(IR)=M
106 W=AI(K,1)
      NHK1=NN-1
      DO 108 J=1,NHK1
108 AI(K,J)=AI(K,J+1)/W
      AI(K,NN)=1.0/W
      DO 109 I=1,NN
      IF(I.EQ.K) GO TO 109
      W=AI(I,1)
      NHK2=NN-1
      DO 110 J=1,NHK2
110 AI(I,J)=AI(I,J+1)-W*AI(K,J)
      AI(I,NN)=-W*AI(K,NN)
109 CONTINUE
103 CONTINUE
      NHK3=NN-1
      DO 111 K=1,NHK3
      IF(K.EQ.IND(K)) GO TO 111
      NHK4=K+1
      DO 112 I=NHK4,NN
      IF(K.NE.IND(I)) GO TO 112
      IR=I
      GO TO 114
112 CONTINUE
114 DO 115 J=1,NN
      W=AI(J,K)
      AI(J,K)=AI(J,IR)
115 AI(J,IR)=W
      IND(IR)=IND(K)
      IND(K)=K
111 CONTINUE
      RETURN
      END
```

```
C***************************************************
      subroutine CLEANV(M,V)
C     Initializing the vector
      DIMENSION V(M)
      DO 10 I=1,M
         V(I)=0.0
      10 CONTINUE
      RETURN
      END
C***************************************************
      SUBROUTINE PROVEM(M,N,AI,B,X,MS,NS)
C     Obtain vector X(M) by
C     taking product of matrix A(M,N) and B(N)
C     However array declaration is AI(MS,NS),B(NS),X(MS)
      DIMENSION AI(MS,NS),B(NS),X(MS)
      DO 10 I=1,M
         X(I)=0.0
         DO 20 J=1,N
            X(I)=AI(I,J)*B(J)+X(I)
   20 CONTINUE
   10 CONTINUE
      RETURN
      END
C***************************************************
      subroutine equal(ms,ns,m,n,x,y)
C     Copy matrix x(m,n)<-y(m,n)
C     However array declaration is x(ms,ns),y(ms,ns)
      dimension x(ms,ns),y(ms,ns)
      do i=1,m
         do j=1,n
            x(i,j) = y(i,j)
         enddo
      enddo
      return
      end
```

Certain aspects of the above program are noteworthy. The subroutines (excluding CLEAN and CLEANV) import integers n and m as well as the array declaration variables at the beginning of the main section (integers ns and ms) as separate arguments defining the array size of the vectors and matrices. In this program, $ns = n = 14$, giving $ms = m = 2$. However, when designing a program package for multiple problems, the array declarations (ns,ms) in the main section should be kept the larger side and their size should be adapted to the problem of interest.

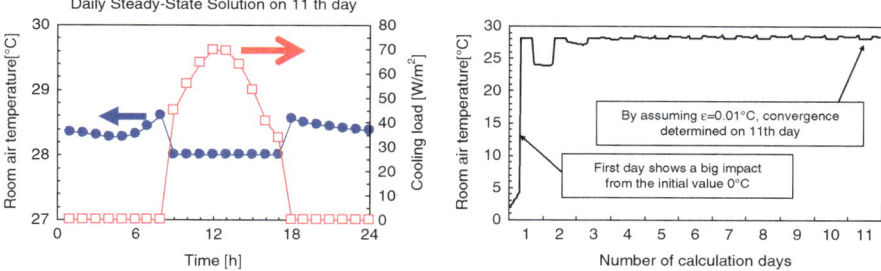

Fig. 2.30 (*Left*) Daily steady-state solution; change over time of room temperature of heat load; (*right*) movements of room temperature leading to daily steady-state solution

The calculated results are output to the file result.csv. Figure 2.30 shows the output from the final day, i.e., the temporal changes of the daily steady-state solutions of room temperature and heat load. The right panel in the figure shows the effect of setting the initial temperature of all nodes to 0 °C. From the changes in room temperature during days 1–11, we observe a periodic steady-state after about 3 days. This example is relevant to small buildings with much smaller heat capacity than soil; thus, large series of calculations are not required to attain daily steady-state solutions.

The reader should reproduce this program and conduct numerous simulations. The following problems are provided as a guide.

{Task 1} How do changes in window size, room depth, and external wall length affect heat load?
{Task 2} What happens to the cooling load if an appropriate level of internal heat is generated?
{Task 3} Investigate the effect of ventilation rate on cooling load.
{Task 4} Obtain the cooling load under 24-h air-conditioning. Consider the obtained cooling load as heat deducted from internal generation and calculate the 24-h natural room temperature to observe the resulting changes in room temperature.
{Task 5} As explained in the footnote of page 38, the above program assumes Eq. (2.73.4). In the time step in which the air-conditioning is switched from off to on, Eq. (2.73.3) should ideally be applied. What changes to the matrix will implement this correction? Moreover, implement these changes in the code.

2.14 Finite Element Method

In this section, we introduce the FEM, a spatial discretization method that differs from the CVM adopted so far.

First, we insist that FEM spatial discretization does not alter the system state Eq. (2.18). As we have reiterated many times, the representation of Eq. (2.18) is

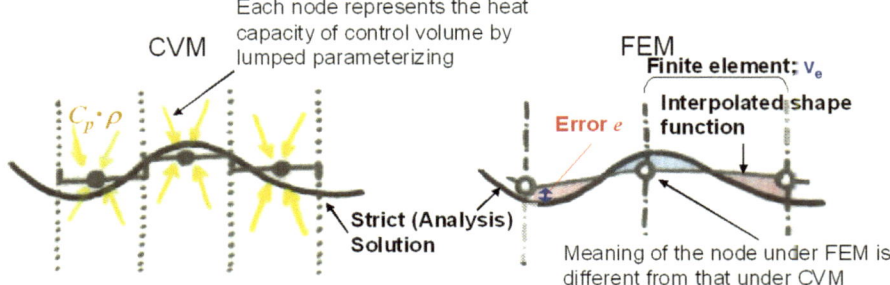

Fig. 2.31 Conceptual difference in space discretization between control volume method and finite element method

universal. However, the vector and matrix constructs differ between CVM and FEM. These differences are most clearly seen in the heat capacitance matrix. The reader should not be put off by the frequent appearance of numerical equations in this section; the content is not difficult and should be perused without fearing that the mathematics will become intractable.

The concept of space discretization using the CVM is illustrated in the left panel of Fig. 2.31.

Suppose that the exact solution of Eq. (2.1), given by the original continuous system, is the solid line in Fig. 2.31. In CVM, the heat capacity within the control volume is lumped parametrized on the central nodes, and the heat balance equation at a node (2.1) is solved by first order integration in the control volume (for example, Eqs. (2.8)–(2.17)). Solutions at all nodes are obtained by solving the simultaneous equations for the whole system. The heat balance within the total control volume satisfies the original mathematical model equation (2.1), but this simply declares that the temperature everywhere within the volume element is that at the node. Because the temperature distribution between the nodes is not considered within a volume, solutions may become discontinuous at the control volume boundaries, as shown in the figure (although of course, if the discretization widths are sufficiently small, a virtually continuous temperature distribution is obtained). In fact, the distribution is frequently obtained by a line joining the node temperatures dot-to-dot. Evidently, a finite control volume will always introduce an error (space discretization error) in the numerical solution of Eq. (2.1).

FEM works on a completely different principle. In a manner of speaking, one could suggest that FEM is much more sophisticated than CVM. First, the meaning of a node in FEM is fundamentally different from that in other space discretization methods. The implications of lumped parametrizing in the representation of heat capacity are absent (hence, there are no distinctions between ○ and ● in the FEM). Initially, the region is divided into finite sized elements, V_e. In one dimension, the nodes are placed on neighboring elements, implying that "the boundary points are specifically named." At this point, suppose we wish to obtain the temperature at an arbitrary position between two nodes by some method (in reality, by approximate interpolation). Then it should be possible to evaluate the error e between the analytical solution and the interpolated temperature at the arbitrary point on the basis of FEM. Ideally, this error should be zero, but zero error cannot be achieved in practice, because the numerical and analytical solutions would then be identical. As the next best solution, the node temperatures at both edges are set such that zero error is obtained by space integrating within the finite element. This principle underlies the *Rayleigh-Ritz-Galerkin Method*, the fundamental FEM approach. Assigning a weight function w to the error at the arbitrary position e, the aforementioned idea is expressed as

$$\int_{v_e} (w \cdot e) dv = 0, \qquad (2.89)$$

where the error is

$$e = [(\text{Exact Solution}) - (\text{Numerical Solution})]_{\text{Arbitrary Position within Finite Element}}.$$
$$(2.90)$$

If the numerical solution at time t at the optional position x within the limited elements is given by $\theta_N(x,t)$ and the exact solution is θ, Eq. (2.90) becomes

$$e = \theta - \theta_N(x,t) = C_p\rho\frac{\partial\theta_N(x,t)}{\partial t} - \lambda\frac{\partial^2\theta_N(x,t)}{\partial x^2}. \qquad (2.91)$$

If $\theta_N(x,t) = \theta$ in the above equation, the error is $e = 0$ from Eq. (2.1) and the correctness of this expression can be appreciated.

Now we must estimate $\theta_N(x,t)$. The temperatures at the extreme nodes of the finite elements are expressed numerically as $\Theta(t)$. The temperature within the element $\theta_N(x,t)$ is obtained by interpolating between the two extreme node temperatures $[\Theta(t)]$. Defining the interpolation function by $[N(x)]$, we obtain

$$\theta_N(x,t) = [N(x)](\Theta(t)). \qquad (2.92)$$

In FEM, this interpolation function is referred to as the shape function. If $[N(x)]$ is adopted as the weight function of Eq. (2.89) (since the weight function can be arbitrarily selected), Eqs. (2.91) and (2.92) can be substituted into Eq. (2.89) to yield

$$\int_{v_e} (N \cdot e) dv = \int_{v_e} {}^T[N] \left(C_p\rho\frac{\partial\theta_N(x,t)}{\partial t} - \lambda\frac{\partial^2\theta_N(x,t)}{\partial x^2} \right) dv = 0. \qquad (2.93)$$

Here ${}^T[N]$ denotes the transposed matrix of $[N(x)]$. This completes the FEM grid setup.

We now derive the discretization equation, assuming the setup of Fig. 2.32. This example is similar to that of Fig. 2.13. The volume is discretized into four finite elements [1]–[4] delineated by five nodes.

Fig. 2.32 Application of
FEM to the analysis of
single-layer wall with
convection heat transfer
boundaries on both sides

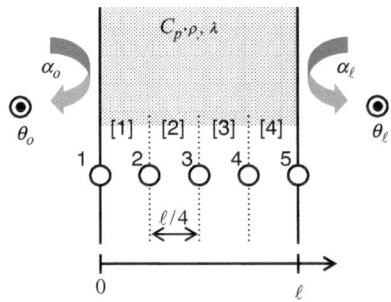

Fig. 2.33 Approximation
within element in the local
coordinate system based on
1-D function

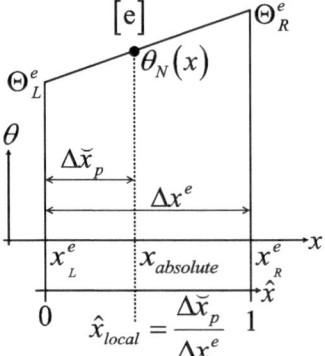

First, we find an explicit form of the shape function. The simplest approximation to the temperature between two edge nodes is a linear interpolation:

$$\theta_N(x, t) \equiv a_1 + a_2 \cdot x. \tag{2.94}$$

Smoother, more accurate interpolations are possible using isoparametric elements[10] but here the simplest functional interpolation will suffice. Consider the finite element [e] in Fig. 2.33. The absolute coordinates for the right and left edge nodes are given by x_L^e and x_R^e, respectively, with respective node temperatures Θ_L^e and Θ_R^e. From Eq. (2.94), we have

$$\begin{cases} \Theta_L^e = a_1 + a_2 \cdot x_L^e \\ \Theta_R^e = a_1 + a_2 \cdot x_R^e \end{cases}. \tag{2.95}$$

---

[10] Despite this, Eq. (2.92), as a general expression, interpolates using the highest linear function of the temperature at the edge nodes in the finite element. "Smooth" is within that possible in a linear approximation.

Solving these for a_1 and a_2 and substituting into Eq. (2.94), the shape function is obtained as

$$\theta_N(x) = \frac{x_R^e - x}{x_R^e - x_L^e}\Theta_L^e + \frac{x - x_L^e}{x_R^e - x_L^e}\Theta_R^e \equiv N_1(x) \cdot \Theta_L^e + N_2(x) \cdot \Theta_R^e. \tag{2.96}$$

We now introduce the local coordinate system $\tilde{x}$ shown in the Fig. 2.33. This coordinate system $\tilde{x}$ is standardized (normalized) with the left edge 0 and the right edge 1. In terms of the absolute coordinates, $\tilde{x} \equiv \frac{x - x_L^e}{x_R^e - x_L^e}$. In the local coordinate system, the shape functions $N_1(x)$ and $N_2(x)$ are

$$\begin{cases} N_1(x) = 1 - \tilde{x} \equiv N_1(\tilde{x}) \\ N_2(x) = \tilde{x} \equiv N_2(\tilde{x}) \end{cases}. \tag{2.97}$$

We now require the scale ratio between the absolute and local coordinates. From the sizes of the finite element [e] in each coordinate system, we obtain

$$\frac{\ell}{4} : 1 = dx : d\tilde{x} \quad \Leftrightarrow \quad dx = \frac{\ell}{4}d\tilde{x}. \tag{2.98}$$

Equation (2.98) expresses the length ratio between the absolute and local coordinates, and thus plays the role of the Jacobian, familiar from concepts such as change of variables in multiple integrals.

Equation (2.93) can be re-written as

$$C_p\rho \int_{v_e} {}^T[N] \frac{\partial \theta_N(x,t)}{\partial t} dv - \lambda \int_{v_e} {}^T[N] \frac{\partial^2 \theta_N(x,t)}{\partial x^2} dv = 0. \tag{2.99}$$

Each term on the left side of Eq. (2.99) will be further transformed as shown below. First, we apply Gauss' divergence theorem to the second term on the left side to yield

$$= -\lambda \int_{v_e} \frac{\partial^T[N]}{\partial x} \frac{\partial \theta_N(x,t)}{\partial x} dv + \lambda \int_{s_e} {}^T[N] \frac{\partial \theta_N(x,t)}{\partial x} ds$$

$$= -\lambda \int_{v_e} \frac{\partial^T[N]}{\partial x} \frac{\partial [N]}{\partial x} dv(\Theta) + \lambda \int_{s_e} {}^T[N] \frac{\partial \Theta}{\partial x} ds. \tag{2.100}$$

Here Gauss' divergence theorem[11] is used to obtain the right side of the first equals sign. The right side of the second equals sign is obtained from Eq. (2.92). v_e and s_e denote the finite element volume of element [e] and its boundary, respectively. Now, in the boundary integral in Eq. (2.100) (the second term on the right- most side), we need to only consider the elements touching the system boundary (elements [1] and [4] in Fig. 2.32). The first term on the right-most side involves the finite elements with no boundaries (elements [2] and [3]). The boundary surface between elements [1] and [4] establishes the following boundary condition:

(flux propagated by conduction) = (flux propagated by convection).

Mathematically, this is expressed as

$$-\lambda \frac{\partial \Theta}{\partial x}\bigg|_{s_1} = \alpha_0(\Theta_1 - \theta_o), \quad -\lambda \frac{\partial \Theta}{\partial x}\bigg|_{s_4} = \alpha_0(\Theta_5 - \theta_\ell). \qquad (2.101)$$

Figure 2.34 Gauss divergence theorem. Illustrating the above, the explicit forms of the right side of Eq. (2.100) for each element are as follows:

---

[11] Gauss Divergence Theorem

The divergence theorem states that the integration over volume V of the divergence of vector $\mathbf{u}$ is equivalent to surface integration of the normal component of $\mathbf{u}$ over the boundary curve S surrounding V (this should make sense physically). Figure 2.34 shows this. Mathematically, this is expressed

$$\int_V div\mathbf{u}dV \left(= \int_V \nabla \bullet \mathbf{u}dV \right) = \int_S \mathbf{u} \bullet \mathbf{n}dS$$

$$\Leftrightarrow \int_V \frac{\partial u_i}{\partial x_i}dV = \int_S u_i \cdot n_i dS,$$

where $div\mathbf{u} = \frac{\partial u_x}{\partial x} + \frac{\partial u_y}{\partial y} + \frac{\partial u_z}{\partial z} = \nabla \bullet \mathbf{u}$.

Substituting $u \equiv vw$, Gauss' divergence theorem is expressed as

$$\int_V \frac{\partial}{\partial x_i}(vw)_i dV = \int_S (vw)_i \cdot n_i dS$$

$$\Leftrightarrow \int_V \frac{\partial v_i}{\partial x_i} w_i dV = \int_S (vw)_i \cdot n_i dS - \int_V v_i \frac{\partial w_i}{\partial x_i} dV.$$

On the right side of the equivalence sign, the formula for the derivative of an integral, $(f \cdot g)' = f' \cdot g + f \cdot g'$ is used. The partial integration formula learned at senior high school, $\int (f \cdot g)' = \int (f' \cdot g + f \cdot g') \Leftrightarrow \int f \cdot g' = f \cdot g - \int f' \cdot g$, is basically equivalent to Gauss' divergence theorem.

Fig. 2.34 Gauss
divergence theorem

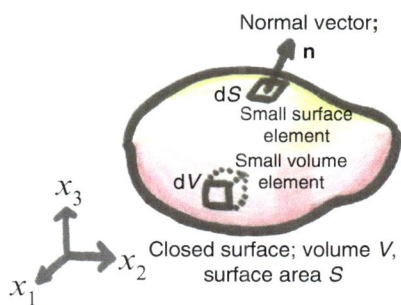

Normal vector; **n**

$\text{d}S$ Small surface element

Small volume element

$\text{d}V$

x_3

x_2

x_1

Closed surface; volume V, surface area S

$$\text{Element}\,[1] = \left[-\lambda \int_{v_1} \frac{\partial^T[N]}{\partial x}\frac{\partial[N]}{\partial x}dx - \alpha_o \int_{s_1} {}^T[N][N]dx\right] \begin{pmatrix} \Theta_1 \\ \Theta_2 \end{pmatrix}$$

$$+\,\alpha_o\theta_o \int_{s_1} {}^T[N]dx, \tag{2.102}$$

$$\text{Element}\,[2] = \left[-\lambda \int_{v_2} \frac{\partial^T[N]}{\partial x}\frac{\partial[N]}{\partial x}dx\right] \begin{pmatrix} \Theta_2 \\ \Theta_3 \end{pmatrix}, \tag{2.103}$$

$$\text{Element}\,[3] = \left[-\lambda \int_{v_3} \frac{\partial^T[N]}{\partial x}\frac{\partial[N]}{\partial x}dx\right] \begin{pmatrix} \Theta_3 \\ \Theta_4 \end{pmatrix}, \tag{2.104}$$

$$\text{Element}\,[4] = \left[-\lambda \int_{v_4} \frac{\partial^T[N]}{\partial x}\frac{\partial[N]}{\partial x}dx - \alpha_\ell \int_{s_4}^T [N][N]dx\right] \begin{pmatrix} \Theta_4 \\ \Theta_5 \end{pmatrix}$$

$$+\,\alpha_\ell\theta_\ell \int_{s_4}^T [N]dx. \tag{2.105}$$

From Fig. 2.32, the term $\alpha_o \int_{s_1} {}^T[N][N]dx$ in Eq. (2.102) involves only a left

boundary; hence, $N_2 = 0$ in the shape function (2.97). Similarly $\alpha_\ell \int_{s_4}^T [N][N]dx$ in

Eq. (2.105) involves only a right boundary, yielding $N_1 = 0$ in the shape function.
   We now consider the first term on the left side of Eq. (2.99). Applying Eq. (2.92)
yet again, we obtain;

$$(\text{1st term on the left side of equation (2.99)}) = C_p\rho \int_{v_e} {}^T[N][N]dv\,\frac{\partial}{\partial t}(\Theta). \tag{2.106}$$

Writing the first term on the left of Eq. (2.99) explicitly for each element, we get

$$\text{Element}\,[1] = C_p\rho \int_{v_1}^{T} [N][N]dv \frac{\partial}{\partial t}\begin{pmatrix}\Theta_1 \\ \Theta_2\end{pmatrix} \equiv [m_1]\frac{\partial}{\partial t}\begin{pmatrix}\Theta_1 \\ \Theta_2\end{pmatrix}, \qquad (2.107)$$

$$\text{Element}\,[2] = C_p\rho \int_{v_2}^{T} [N][N]dv \frac{\partial}{\partial t}\begin{pmatrix}\Theta_2 \\ \Theta_3\end{pmatrix} \equiv [m_2]\frac{\partial}{\partial t}\begin{pmatrix}\Theta_2 \\ \Theta_3\end{pmatrix}, \qquad (2.108)$$

$$\text{Element}\,[3] = C_p\rho \int_{v_3}^{T} [N][N]dv \frac{\partial}{\partial t}\begin{pmatrix}\Theta_3 \\ \Theta_4\end{pmatrix} \equiv [m_3]\frac{\partial}{\partial t}\begin{pmatrix}\Theta_3 \\ \Theta_4\end{pmatrix}, \qquad (2.109)$$

$$\text{Element}\,[4] = C_p\rho \int_{v_4}^{T} [N][N]dv \frac{\partial}{\partial t}\begin{pmatrix}\Theta_4 \\ \Theta_5\end{pmatrix} \equiv [m_4]\frac{\partial}{\partial t}\begin{pmatrix}\Theta_4 \\ \Theta_5\end{pmatrix}. \qquad (2.110)$$

The above step completes the FEM construction. Equation (2.99) has been expressed as a function of each node temperature enclosing each finite element. To express Eq. (2.99) as a system state equation, the above four equations are combined into matrix form. In this example, because the boundary condition vector **f** of the heat flux input contains all zeros, the system state equation becomes

$$\mathbf{M}\frac{d\mathbf{\Theta}}{dt} = \mathbf{C}\mathbf{\Theta} + \mathbf{C_o}\mathbf{\Theta_o}. \qquad (2.111)$$

Here $^T\mathbf{\Theta} = [\,\Theta_1 \quad \Theta_2 \quad \Theta_3 \quad \Theta_4 \quad \Theta_5\,]$ is an unknown variable vector. In the following analysis, Eqs. (2.102)–(2.105) and (2.107)–(2.110) are summarized and each vector and matrix element in Eq. (2.111) is explicitly written.

First, the vector matrix product $\mathbf{C_o}\mathbf{\Theta_o}$ describing the boundary condition of convection is

$$\mathbf{C_o}\mathbf{\Theta_o} = \begin{bmatrix}\ \\ \ \\ \ \end{bmatrix} \begin{cases} \begin{bmatrix}\ \end{bmatrix} = \left[\alpha_o\theta_o\int_{s_1}^{T}[N]dx\right] = \alpha_o\theta_o\begin{bmatrix}N_1 \\ N_2\end{bmatrix}_{s_1} = \alpha_o\theta_o\begin{bmatrix}1 \\ 0\end{bmatrix} \\[18pt] \begin{bmatrix}\ \end{bmatrix} = \left[\alpha_\ell\theta_\ell\int_{s_4}^{T}[N]dx\right] = \alpha_\ell\theta_\ell\begin{bmatrix}N_1 \\ N_2\end{bmatrix}_{s_4} = \alpha_\ell\theta_\ell\begin{bmatrix}0 \\ 1\end{bmatrix} \end{cases} \qquad (2.112)$$

$$= {}^T[\alpha_o\theta_o \quad 0 \quad 0 \quad 0 \quad \alpha_\ell\theta_\ell]$$

Thus, Eqs. (2.101) and (2.104), denoting the second term in Eq. (2.99), become

$$\mathbf{C}\mathbf{\Theta} = \boxed{}\ \mathbf{\Theta}$$

$$(2.113.1)$$

$$\boldsymbol{\Theta} = \left[-\lambda \int_{v_1} \frac{\partial^T [N]}{\partial x} \frac{\partial [N]}{\partial x} dx - \alpha_o \int_{s_1}^T [N][N] dx \right] \begin{pmatrix} \Theta_1 \\ \Theta_2 \end{pmatrix}$$

$$= \left[\lambda \frac{4}{\ell} \begin{bmatrix} -1 & 1 \\ 1 & -1 \end{bmatrix} - \alpha_o \int_{s_1} \begin{bmatrix} 1 \\ 0 \end{bmatrix} [1 \quad 0] dx \right] \begin{pmatrix} \Theta_1 \\ \Theta_2 \end{pmatrix}$$

$$= \lambda \frac{4}{\ell} \begin{bmatrix} -1 - \alpha_o \dfrac{\ell}{4\lambda} & 1 \\ 1 & -1 \end{bmatrix} \begin{pmatrix} \Theta_1 \\ \Theta_2 \end{pmatrix}, \qquad (2.113.2)$$

$$\boldsymbol{\Theta} = \left[-\lambda \int_{v_2} \frac{\partial^T [N]}{\partial x} \frac{\partial [N]}{\partial x} dx \right] \begin{pmatrix} \Theta_2 \\ \Theta_3 \end{pmatrix} = -\lambda \int_0^1 \frac{d\tilde{x}}{dx} \frac{\partial^T [N]}{\partial \tilde{x}} \frac{d\tilde{x}}{dx} \frac{\partial [N]}{\partial \tilde{x}} \frac{\ell}{4} d\tilde{x} \begin{pmatrix} \Theta_2 \\ \Theta_3 \end{pmatrix}$$

$$= -\lambda \frac{4}{\ell} \int_0^1 \frac{\partial^T [N]}{\partial \tilde{x}} \frac{d\tilde{x}}{dx} \frac{\partial [N]}{\partial \tilde{x}} d\tilde{x} \begin{pmatrix} \Theta_2 \\ \Theta_3 \end{pmatrix} = -\lambda \frac{4}{\ell} \int_0^1 \begin{bmatrix} \dfrac{\partial N_1}{\partial \tilde{x}} \\ \dfrac{\partial N_2}{\partial \tilde{x}} \end{bmatrix} \begin{bmatrix} \dfrac{\partial N_1}{\partial \tilde{x}} & \dfrac{\partial N_2}{\partial \tilde{x}} \end{bmatrix} d\tilde{x} \begin{pmatrix} \Theta_2 \\ \Theta_3 \end{pmatrix}$$

$$= -\lambda \frac{4}{\ell} \int_0^1 \begin{bmatrix} -1 \\ 1 \end{bmatrix} [-1 \quad 1] d\tilde{x} \begin{pmatrix} \Theta_2 \\ \Theta_3 \end{pmatrix} = \lambda \frac{4}{\ell} \begin{bmatrix} -1 & 1 \\ 1 & -1 \end{bmatrix} \begin{pmatrix} \Theta_2 \\ \Theta_3 \end{pmatrix},$$

$$(2.113.3)$$

$$\boldsymbol{\Theta} = \lambda \frac{4}{\ell} \begin{bmatrix} -1 & 1 \\ 1 & -1 \end{bmatrix} \begin{pmatrix} \Theta_3 \\ \Theta_4 \end{pmatrix}, \qquad (2.113.4)$$

$$\boldsymbol{\Theta} = \lambda \frac{4}{\ell} \begin{bmatrix} -1 & 1 \\ 1 & -1 - \alpha_\ell \dfrac{\ell}{4\lambda} \end{bmatrix} \begin{pmatrix} \Theta_4 \\ \Theta_5 \end{pmatrix}, \qquad (2.113.5)$$

$$\mathbf{M} \frac{d\boldsymbol{\Theta}}{dt} = \boxed{} \frac{d\boldsymbol{\Theta}}{dt}$$

$$(2.114.1)$$

However, for element $[i]$ we have

$$[m_I] \frac{d\boldsymbol{\Theta}}{dt} = C_p \rho \int_{v_i}^T [N][N] dv \frac{\partial}{\partial t} \begin{pmatrix} \Theta_i \\ \Theta_{i+1} \end{pmatrix}$$

$$= C_p \rho \int_0^1 \begin{bmatrix} 1 - \tilde{x} \\ \tilde{x} \end{bmatrix} [1 - \tilde{x} \quad \tilde{x}] \frac{\ell}{4} d\tilde{x} \frac{\partial}{\partial t} \begin{pmatrix} \Theta_i \\ \Theta_{i+1} \end{pmatrix}$$

$$= \frac{C_p \rho \ell}{4} \int_0^1 \begin{bmatrix} 1 - 2\tilde{x} + \tilde{x}^2 & \tilde{x} - \tilde{x}^2 \\ \tilde{x} - \tilde{x}^2 & \tilde{x}^2 \end{bmatrix} d\tilde{x} \; \frac{\partial}{\partial t} \begin{pmatrix} \Theta_i \\ \Theta_{i+1} \end{pmatrix}$$

$$= \frac{C_p \rho \ell}{4} \begin{bmatrix} \int_0^1 (1 - 2x + x^2) dx & \int_0^1 (x - x^2) dx \\ \int_0^1 (x - x^2) dx & \int_0^1 x^2 dx \end{bmatrix} \frac{\partial}{\partial t} \begin{pmatrix} \Theta_i \\ \Theta_{i+1} \end{pmatrix}$$

$$= \frac{C_p \rho \ell}{24} \begin{bmatrix} 2 & 1 \\ 1 & 2 \end{bmatrix} \frac{\partial}{\partial t} \begin{pmatrix} \Theta_i \\ \Theta_{i+1} \end{pmatrix}.$$

$$(2.114.2)$$

Note that in Eq. (2.114.2), the sum of all elements of $\frac{\ell}{24} \begin{bmatrix} 2 & 1 \\ 1 & 2 \end{bmatrix}$ is $\frac{\ell}{24}(2 + 1 + 1 + 2) = \frac{\ell}{4}$, the volume of the finite element. In other words, the matrix **M** in the CVM includes the heat capacity of the entire control volume in its diagonal elements. By contrast, in the FEM, heat capacity is distributed among the 2×2 elements around the adjacent two nodes.

Finally, we highlight the differences between the vectors and matrices of the system state Eq. (2.110) formulated in CVM and FEM. As an illustrative example, we consider space discretization using a 5-node CVM with no surface heat capacity (as in Fig. 2.32; see Fig. 2.35). In this 5-node model, the **C** matrix of CVM differs from that of FEM because CVM includes surface nodes with no heat capacity. On the other hand, as discussed above, the **M** matrix is fundamentally different between the two approaches. Although the CVM formulation imposes diagonal elements because of lumped parameterization, FEM produces a band matrix with non-diagonal elements. In this case, as mentioned in Sect. 2.6, because the time discretization scheme is forward difference, the inverse matrix must be computed, and FEM offers no explicit solution. Moreover, if the stability condition for the numerical solution is imposed, the FEM solution confers no advantages.

2.15   End of Chapter Examples

This section will solidify (in the readers' mind) the reasoning behind the system state equations introduced so far, through a set of practical examples. Each example involves the explicit expression of vectors and matrices in the system state Eq. (2.18).

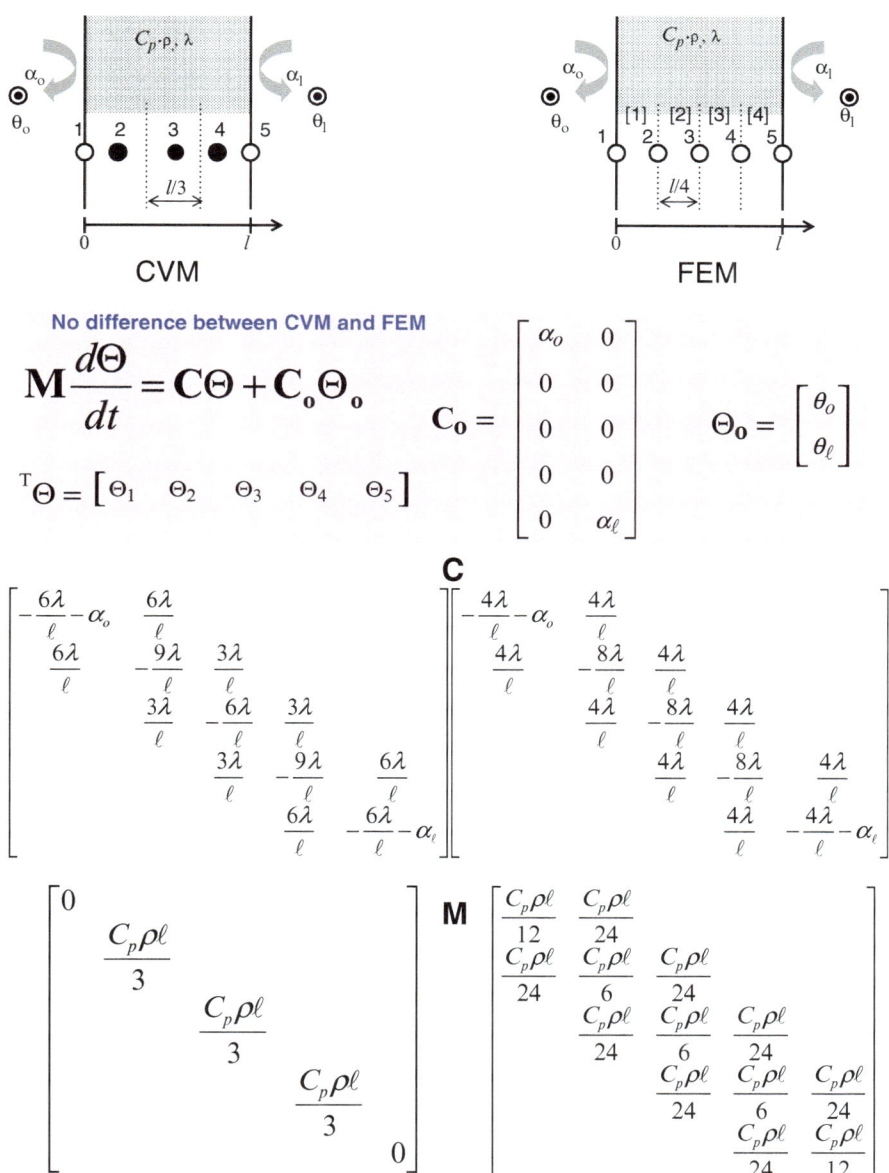

Fig. 2.35 Difference between matrices **C** and **M** in the CVM and FEM formulations of the same system state equations

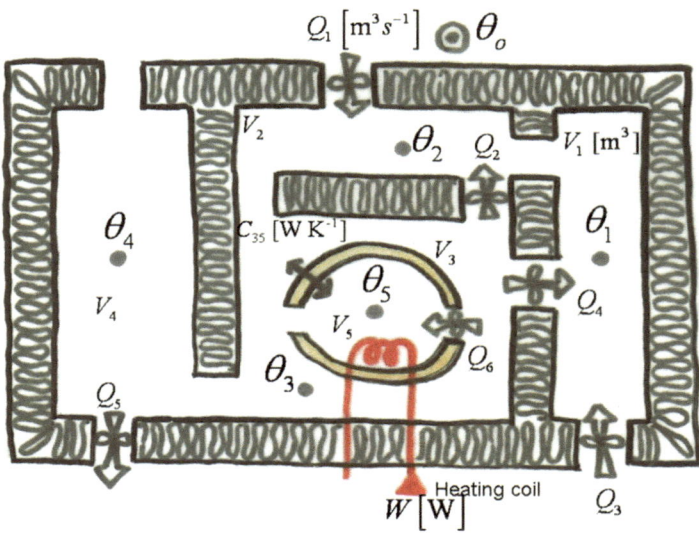

Fig. 2.36 Heat system in question 1

Example 1 Consider a heat system comprising 5 rooms, as shown in Fig. 2.36. Room 5 is enclosed by room 3 and is ventilated by the air from room 3 by a fan Q_6 [m³/s]. At other openings, air is forcefully fan-ventilated in directions shown by the arrows. Room 5 is heated by W [W]. The conductance between rooms 5 and 3 is given by C_{35} [W/K] (note that this quantity already contains the surface area's influence). Other walls are assumed perfectly insulated (as shown in the figure) and the heat transfer between the wall surfaces and the room temperature nodes can be ignored. Moreover, the relationship $Q_1 + Q_3 > Q_5$ holds.

Solution The unknown temperature node vector is defined by $\boldsymbol{\theta} = {}^T[\theta_1 \quad \cdots \quad \theta_5]$. Because the total heat flow in each room is zero and $Q_1 + Q_3 > Q_5$, the magnitude and direction of heat flow at each opening surface is determined as shown in Fig. 2.37.

In this situation, the vectors and matrices of Eq. (2.18) $\mathbf{M}\frac{d\boldsymbol{\theta}}{dt} = \mathbf{C}\boldsymbol{\theta} + \mathbf{C_o}\boldsymbol{\theta_o} + \mathbf{f}$ are expressed as

$$\mathbf{M} = \left(C_p\rho\right)_{air} \begin{bmatrix} V_1 & & & & \\ & V_2 & & & \\ & & V_3 & & \\ & & & V_4 & \\ & & & & V_5 \end{bmatrix}$$

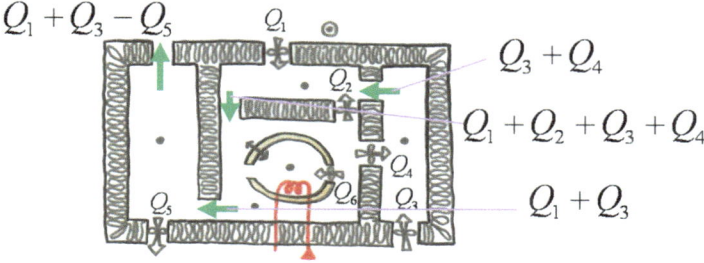

Fig. 2.37 Heat flow at each opening in Example 1

$$\mathbf{C} = (C_p\rho)_{air}$$

$$\begin{bmatrix} -(Q_3 + Q_4) & & Q_4 & & \\ (Q_3 + Q_4) & -(Q_1 + Q_2 + Q_3 + Q_4) & Q_1 & & \\ & (Q_1 + Q_2 + Q_3 + Q_4) & -(Q_1 + Q_2 + Q_3 + Q_4 + Q_6) & & Q_6 \\ & & (Q_1 + Q_3) & -(Q_1 + Q_3) & \\ & & Q_6 & & -Q_6 \end{bmatrix}$$

$$\mathbf{C_o} = \begin{bmatrix} Q_3 (C_p\rho)_{air} \\ Q_1 (C_p\rho)_{air} \\ \\ \end{bmatrix} \quad \boldsymbol{\theta_o} = [\theta_o] \quad \mathbf{f} = \begin{bmatrix} \\ \\ \\ W \end{bmatrix}$$

Example 2 Consider a heat system comprising four rooms plus an air conditioning room, as shown in Fig. 2.38. The magnitudes and units of each variable are those of Example 1. In this situation, the air-conditioning room resides upstream of room 4. External air introduced to room 4 is adjusted to $\theta_a°C$ by cooling and heating coils*.

Solution The unknown temperature node vector is defined as $\boldsymbol{\theta} = {}^T[\theta_1 \quad \theta_2 \quad \theta_3 \quad \theta_4]$. To maintain zero total heat flow into each room, the magnitude and direction of heat flow at each opening is determined as shown in Fig. 2.39. To model this situation, the vectors and matrices are given by

$$\mathbf{M} = (C_p\rho)_{air} \begin{bmatrix} V_1 & & & \\ & V_2 & & \\ & & V_3 & \\ & & & V_4 \end{bmatrix}$$

$$\mathbf{C} = \begin{bmatrix} -(Q_a + Q_b)(C_p\rho)_{air} & Q_b(C_p\rho)_{air} & & \\ & -Q_b(C_p\rho)_{air} - A_g C_{2o} + \varpi & Q_b(C_p\rho)_{air} & \\ & & -Q_c(C_p\rho)_{air} & Q_c(C_p\rho)_{air} \\ & & (Q_c - Q_b)(C_p\rho)_{air} & -Q_c(C_p\rho)_{air} \end{bmatrix}$$

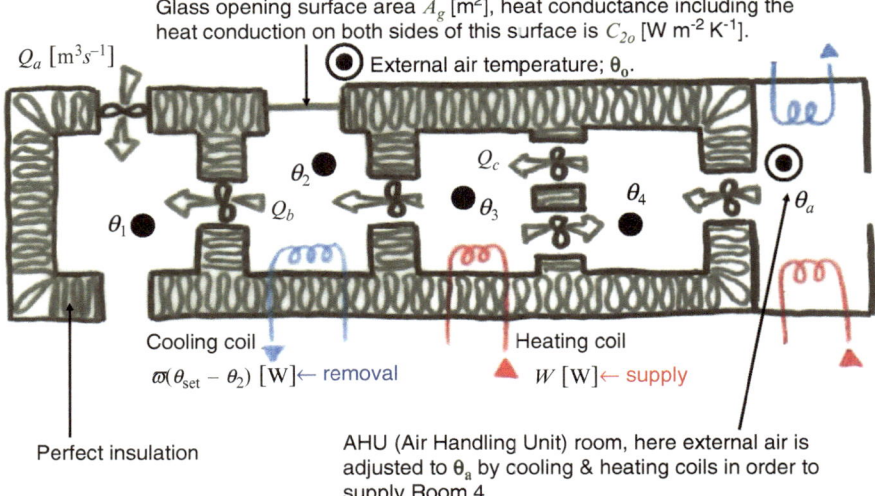

Fig. 2.38 Heat system in Example 2

Fig. 2.39 Heat flow at each
opening in question 1

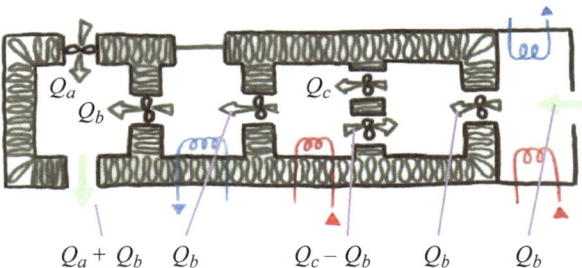

$$\mathbf{C_o} = \begin{bmatrix} Q_a(C_p\rho)_{air} \\ A_gC_{2o} \\ & & -\varpi \\ & Q_b(C_p\rho)_{air} \end{bmatrix} \quad \mathbf{\theta_o} = \begin{bmatrix} \theta_o \\ \theta_a \\ \theta_{set} \end{bmatrix} \quad \mathbf{f} = \begin{bmatrix} W \end{bmatrix}$$

Example 3 Consider a heat system structured from four rooms, as shown in
Fig. 2.40. Internal heats $W_1 - W_3$ are generated in rooms 1–3, while room 4 is
cooled to θ_{set}°C by air-conditioning. The heat load is H_{ex} [W]. Furthermore, we
assume that $Q_3 + Q_4 > Q_1$.

Solution The   unknown   temperature   node   vector   is   defined   as
$\mathbf{\theta} = {}^T[\theta_1 \quad \theta_2 \quad \theta_3 \quad H_{ex}]$. Because the total heat flow into each room is zero and
$Q_3 + Q_4 > Q_1$, the magnitudes and directions of heat flow at each opening are
determined as shown in Fig. 2.41. The vector and matrix constructs are
shown below:

Fig. 2.40 Heat system in Example 3

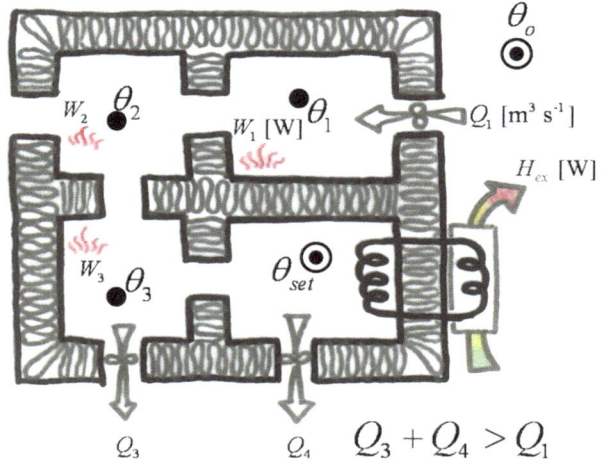

Fig. 2.41 Flow amount at each opening in question 3

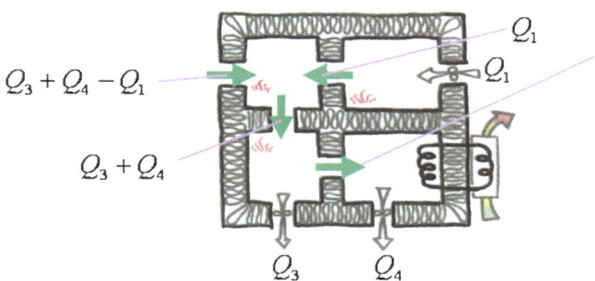

$$\mathbf{M} = \left(C_p\rho\right)_{air} \begin{bmatrix} V_1 & & & \\ & V_2 & & \\ & & V_3 & \\ & & & 0 \end{bmatrix}$$

$$\mathbf{C} = \begin{bmatrix} -Q_1\left(C_p\rho\right)_{air} & & & \\ Q_1\left(C_p\rho\right)_{air} & -(Q_3+Q_4)\left(C_p\rho\right)_{air} & & \\ & (Q_3+Q_4)\left(C_p\rho\right)_{air} & -(Q_3+Q_4)\left(C_p\rho\right)_{air} & \\ & & Q_4\left(C_p\rho\right)_{air} & -1 \end{bmatrix}$$

$$\mathbf{C_o} = \begin{bmatrix} Q_1\left(C_p\rho\right)_{air} \\ (Q_3+Q_4-Q_1)\left(C_p\rho\right)_{air} \\ \\ -Q_4\left(C_p\rho\right)_{air} \end{bmatrix} \quad \boldsymbol{\theta_o} = \begin{bmatrix} \theta_o \\ \theta_{set} \end{bmatrix} \quad \mathbf{f} = \begin{bmatrix} W_1 \\ W_2 \\ W_3 \end{bmatrix}$$

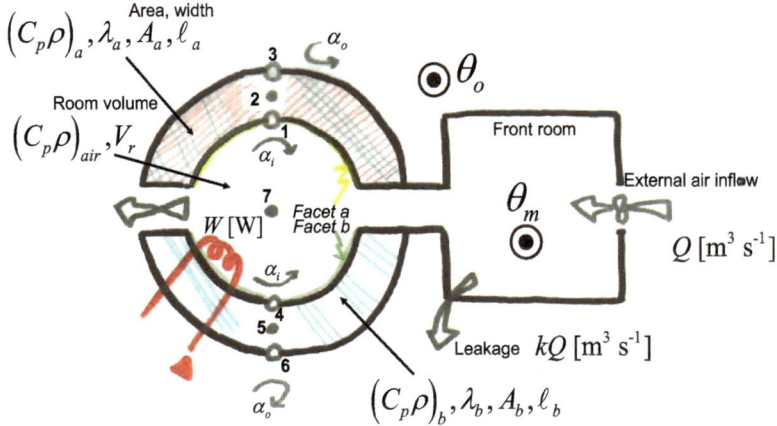

Fig. 2.42 Heat system in Example 4

Example 4 Consider the heat system of Fig. 2.42. External air that enters the front room is adjusted to $\theta_m °C$ and supplied to the downstream room. The front room leaks heat, as shown in the figure. The downstream room comprises facing surfaces a and b, for which the surface to surface form factor between a and b (b and a) is $F_{ab}(F_{ba})$. In addition, surfaces a and b are space discretized and assigned temperatures at their nodes; #1–3 and #4–6. Node #7 indicates room air. Seven temperature nodes are assumed. The physical heat properties of surfaces a and b, wall thickness, and areas, are indicated in the figure. Each surface is backed by a boundary that exchanges heat with external air. Furthermore, the room is supplied with a quantity W [W] of heat.

Solution The unknown temperature node vector is defined by $\boldsymbol{\theta} = {}^T[\theta_1 \quad \cdots \quad \theta_7]$. The vectors and matrices of this problem are constructed as

$$
\mathbf{M} = \begin{bmatrix} 0 \\ & (C_p\rho)_a \ell_a A_a \\ & & 0 \\ & & & 0 \\ & & & & (C_p\rho)_b \ell_b A_b \\ & & & & & 0 \\ & & & & & & (C_p\rho)_{air} V_r \end{bmatrix}
$$

Fig. 2.43 Heat system in
Example 5

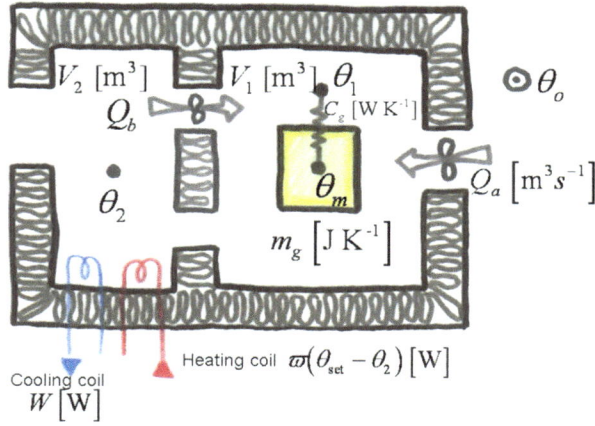

Cooling coil
$W\ [\mathrm{W}]$

Heating coil $\varpi(\theta_{set} - \theta_2)\ [\mathrm{W}]$

$$\mathbf{C} = \begin{bmatrix} -\dfrac{2\lambda_a}{\ell_a} - \alpha_i A_a - \alpha_r A_a F_{ab} & \dfrac{2\lambda_a}{\ell_a} & & & \alpha_r A_a F_{ab} & & \alpha_i A_a \\[2ex] \dfrac{2\lambda_a}{\ell_a} & -\dfrac{4\lambda_a}{\ell_a} & \dfrac{2\lambda_a}{\ell_a} & & & & \\[2ex] & \dfrac{2\lambda_a}{\ell_a} & -\dfrac{2\lambda_a}{\ell_a} - \alpha_o A_a & & & & \\[2ex] \alpha_r A_b F_{ba} & & & -\dfrac{2\lambda_a}{\ell_a} - \alpha_o A_a - \alpha_r A_b F_{ba} & \dfrac{2\lambda_b}{\ell_b} & & \alpha_i A_b \\[2ex] & & & \dfrac{2\lambda_b}{\ell_b} & -\dfrac{4\lambda_b}{\ell_b} & \dfrac{2\lambda_b}{\ell_b} & \\[2ex] & & & & \dfrac{2\lambda_b}{\ell_b} & -\dfrac{2\lambda_b}{\ell_b} - \alpha_o A_b & \\[2ex] \alpha_i A_a & & & \alpha_i A_a & & & -\alpha_i A_a - \alpha_i A_b - (C_p\rho)_{air}(1-k)Q \end{bmatrix}$$

$$\mathbf{C_o} = \begin{bmatrix} \\ \alpha_o A_a \\ \\ \\ \alpha_o A_a \\ (C_p\rho)_{air}(1-k)Q \end{bmatrix} \qquad \mathbf{\theta_o} = \begin{bmatrix} \theta_o \\ \theta_m \end{bmatrix} \qquad \mathbf{f} = \begin{bmatrix} \\ \\ \\ \\ \\ \\ W \end{bmatrix}$$

Example 5 Consider a heat system comprising two rooms as shown in Fig. 2.43. Room 1 contains an object of heat capacity m_g [J/K] whose central temperature θ_m is an unknown quantity in the analysis. The conductance from the lumped parameterized object temperature node θ_m to temperature node θ_1 in room 1 is c_g [W/K] (note that this quantity already contains the surface area's influence). A heating and cooling system is installed in room 2.

Solution The unknown temperature node vector is defined as $\mathbf{\theta} = {}^T[\theta_1 \quad \theta_2 \quad \theta_m]$. Because the heat flow must balance (sum to zero) in each room, the magnitude and

Fig. 2.44 Heat flow at each opening in Example 5

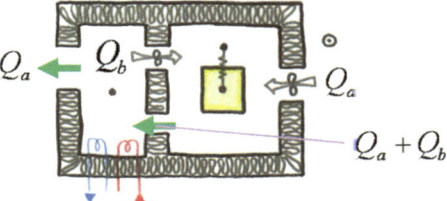

direction of flow at each opening surface is determined as shown in Fig. 2.44. The corresponding vector and matrix constructs are

$$\mathbf{M} = \begin{bmatrix} (C_p\rho)_{air}V_1 & & \\ & (C_p\rho)_{air}V_2 & \\ & & m_g \end{bmatrix}$$

$$\mathbf{C} = \begin{bmatrix} -(Q_a+Q_b)(C_p\rho)_{air}-C_g & Q_b(C_p\rho)_{air} & C_g \\ (Q_a+Q_b)(C_p\rho)_{air} & -(Q_a+Q_b)(C_p\rho)_{air}-\varpi & \\ C_g & & -C_g \end{bmatrix}$$

$$\mathbf{C_o} = \begin{bmatrix} (C_p\rho)_{air}\,Q_a & \\ & \varpi \end{bmatrix} \quad \boldsymbol{\theta_o} = \begin{bmatrix} \theta_o \\ \theta_{set} \end{bmatrix} \quad \mathbf{f} = \begin{bmatrix} -W \end{bmatrix}$$

Example 6 Consider the four-room heat system of Fig. 2.45. As in Example 1, room 4 is enclosed by room 2 and receives air from a fan operating in room 2. The wall separating rooms 2 and 4 comprises facing surfaces $s1$ and $s2$. These surfaces are assumed fully insulated (i.e., transmit no heat). However, the temperature nodes θ_{s1} and θ_{s2} mutually exchange radiant heat and convective heat with room temperature node θ_4. The areas and view factors of surfaces $s1$ and $s2$ are indicated in the figure. In addition, W_3 [W] of heat is generated in room 4. Internal heat W_1 [W] is generated in room 1, which is also air-conditioned to $\vartheta_1°C$. The cooling load in this room (where the amount of heat extracted is taken as positive) is H_{ex} [W]. The wall separating rooms 2 and 3, unlike the walls considered so far, allows heat entry via conduction and surface heat transfer. The thermophysical properties, wall thickness, and surface area are indicated in the figure. In the space discretization, the heat capacity of the entire wall is expressed in terms of the internal temperature node θ_m. Internal heat W_2 [W] is generated in room 3.

Solution The unknown temperature node vector is defined by $\boldsymbol{\theta} = {}^T[H_{ex}\quad \theta_2\quad \theta_3\quad \theta_4\quad \theta_{s1}\quad \theta_{s2}\quad \theta_m]$. To ensure that the heat flow balances in each room, the flow magnitude and direction at each open surface is determined as shown in Fig. 2.46. In this situation, the vector and matrix constructs are

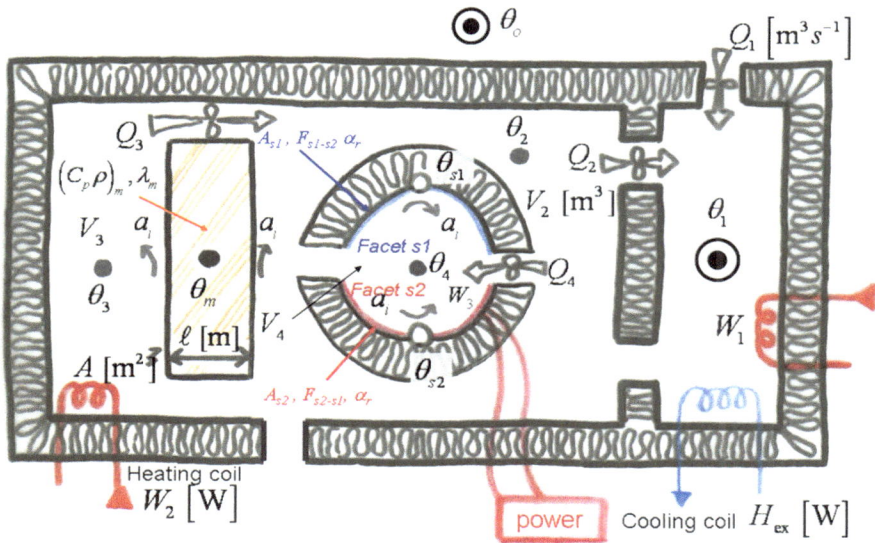

Fig. 2.45   Heat system in Example 6

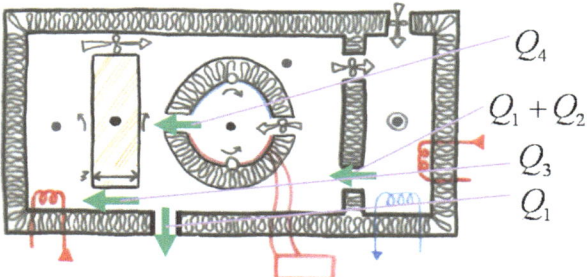

Fig. 2.46   Heat flow at each opening in Example 6

$$\mathbf{M} = \begin{bmatrix} 0 & & & & & \\ & (C_p\rho)_{air} V_2 & & & & \\ & & (C_p\rho)_{air} V_3 & & & \\ & & & (C_p\rho)_{air} V_4 & & \\ & & & & 0 & \\ & & & & & 0 \\ & & & & & & (C_p\rho)_m A\ell \end{bmatrix}$$

Because the heat conductance matrix does not fit on a portrait page, it is shown in landscape configuration on the following page.

$$\mathbf{C_o} = \begin{bmatrix} Q_1(C_p\rho)_{air} & \begin{array}{c} -(Q_1+Q_2)(C_p\rho)_{air} \\ (Q_1+Q_2)(C_p\rho)_{air} \end{array} \\ \\ \\ \end{bmatrix} \qquad \boldsymbol{\theta_o} = \begin{bmatrix} \theta_o \\ \theta_{set} \end{bmatrix} \quad \mathbf{f} = \begin{bmatrix} W_1 \\ W_2 \\ \\ W_3 \end{bmatrix}$$

$$\mathbf{C} = \begin{bmatrix}
-1 & Q_2(C_p\rho)_{air} & & & Q_2(C_p\rho)_{air} & \\[2mm]
& -(Q_1+Q_2+Q_3+Q_4)(C_p\rho)_{air}-\dfrac{A}{\frac{1}{\alpha_i}+\frac{\ell/2}{\lambda_m}} & Q_3(C_p\rho)_{air} & & & \dfrac{A}{\frac{1}{\alpha_i}+\frac{\ell/2}{\lambda_m}} \\[4mm]
& Q_3(C_p\rho)_{air} & -Q_3(C_p\rho)_{air}-\dfrac{A}{\frac{1}{\alpha_i}+\frac{\ell/2}{\lambda_m}} & & & \dfrac{A}{\frac{1}{\alpha_i}+\frac{\ell/2}{\lambda_m}} \\[4mm]
& Q_4(C_p\rho)_{air} & & -Q_4(C_p\rho)_{air}-\alpha_i(A_1+A_2) & \alpha_iA_1 & \alpha_iA_2 \\[2mm]
& \dfrac{A}{\frac{1}{\alpha_i}+\frac{\ell/2}{\lambda_m}} & \dfrac{A}{\frac{1}{\alpha_i}+\frac{\ell/2}{\lambda_m}} & \alpha_iA_1 & -\alpha_iA_1-\alpha_rF_{s2-s1}A_2 & \alpha_rF_{s2-s1}A_2 \\[2mm]
& & & \alpha_iA_2 & \alpha_rF_{s1-s2}A_1 & -\alpha_iA_2-\alpha_rF_{s1-s2}A_1 & -\dfrac{2A}{\frac{1}{\alpha_i}+\frac{\ell/2}{\lambda_m}}
\end{bmatrix}$$

Chapter 3
Applications of Vector Matrix Operations

Abstract In Chap. 2, we introduced the notation of linear system dynamics on the basis of the system state equations. While this notation is extremely useful, its main purpose is to jointly express the governing equations in terms of vectors and matrices. Deviating marginally from the subject, this chapter describes how far you can see into the general system by taking advantage of vectors and matrices, with reference to the *least squares method*.

Keywords Least square solution • Least squares method • Linear multi regression analysis

3.1 Linear Multi Regression Analysis

Many of you may have used linear regression analysis. Let us consider that by taking four environmental factors (temperature, humidity, radiation temperature, and airflow velocity) and two human factors (clothing and metabolic rate) into account, a sensible temperature is defined (such as the SET*). The environment is reproduced in a climatron with these factors varied, and the subjects are asked to answer the thermal sensation vote (TSV) (from −3 (cold) to +3 (hot)). A large amount of experimental data yields a positive correlation between SET* and TSV, and the data can be fitted to the linear regression formula (TSV) = $a \cdot$ (SET*) + b. Such linear regression analysis with a single predictor variable, as in this example, has been generally installed as a standard function in even poor software. Only plotting a graph can immediately provide the regression formula.

  Multi regression analysis, on the other hand, involves several predictor variables. For example, it is assumed that the response variable is energy consumption and that several mutually independent factors such as household income, number of family members, and home location (e.g., latitude) can be suggested as predictor variables. Linear multi regression analysis involves fitting the response variable to the group of predictor variables. In most software and statistical packages, linear multi regression

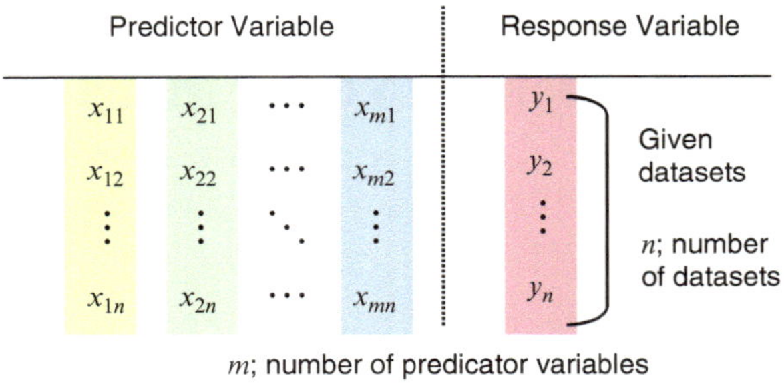

Fig. 3.1 Multi regression
analysis dataset

is conducted with a single click; hence, users may not fully understand the process of the calculation. This section describes the concepts of multi regression analysis and its underlying least squares fitting in vector–matrix format.

Consider a multi regression model with m predictor variables as shown in Fig. 3.1. From left to right, row 1 contains the income of household A, the number of members in the family, latitude of residence, etc. The final column holds the response variable data (namely, energy consumption). Suppose there are n datasets, representing n households.

The multi regression model is:

$$\hat{Y} = b_0 + b_1 X_1 + b_2 X_2 + \cdots + b_m X_m, \tag{3.1}$$

where the X_i's and Y's represent the predictor variables (x_{i*} data) and response variables ($y*$ data), respectively. * is a wild card. $\hat{Y}$ is estimated from the multi regression model and $[b_0, b_1, \cdots b_m]$ are the regression coefficients. We aim to identify the regression coefficient such that the estimated value $\hat{Y}$ from Eq. (3.1) best explains the true value y in the given dataset. "Best explains" means that "the error between the individual estimated and true values is minimized." Note that the errors can be positive or negative. To avoid the problems associated with absolute value calculations, the square of the errors is minimized instead. This approach is known as the *least squares method*. Numerically, we identify $[b_0, b_1, \cdots b_m]$ such that the expression

$$E \equiv \sum_{i=1}^{n} \left(y_i - \hat{Y}_i\right)^2 = \sum_{i=1}^{n} \left(y_i - (b_0 + b_1 x_{1i} + \cdots + b_m x_{mi})\right)^2 \tag{3.2}$$

is minimized. Since E is a function of $[b_0, b_1, \cdots b_m]$, this problem is equivalent to minimizing E; that is

$$\frac{\partial E}{\partial b_0} = 0 \Leftrightarrow -1 \times 2 \sum_{i=1}^{n} (y_i - (b_0 + b_1 x_{1i} + \cdots + b_m x_{mi})) = 0$$

$$\Leftrightarrow n b_0 + b_1 \sum_{i=1}^{n} x_{1i} + b_2 \sum_{i=1}^{n} x_{2i} + b_m \sum_{i=1}^{n} x_{mi} = \sum_{i=1}^{n} y_i \tag{3.3}$$

where the first equivalence is derived from the chain rule of differentiation.

$$\frac{\partial E}{\partial b_1} = 0 \Leftrightarrow -\sum_{i=1}^{n} x_{1i} \times 2\sum_{i=1}^{n}(y_i - (b_0 + b_1 x_{1i} + \cdots + b_m x_{mi})) = 0$$

$$\Leftrightarrow b_0\sum_{i=1}^{n} x_{1i} + b_1\sum_{i=1}^{n} x_{1i}^2 + b_2\sum_{i=1}^{n} x_{1i}x_{2i} + b_m\sum_{i=1}^{n} x_{1i}x_{mi} = \sum_{i=1}^{n} x_{1i}y_i \tag{3.4}$$

$$\frac{\partial E}{\partial b_m} = 0 \Leftrightarrow -\sum_{i=1}^{n} x_{mi} \times 2\sum_{i=1}^{n}(y_i - (b_0 + b_1 x_{1i} + \cdots + b_m x_{mi})) = 0$$

$$\Leftrightarrow b_0\sum_{i=1}^{n} x_{mi} + b_1\sum_{i=1}^{n} x_{1i}x_{mi} + b_2\sum_{i=1}^{n} x_{2i}x_{mi} + b_m\sum_{i=1}^{n} x_{mi}^2 = \sum_{i=1}^{n} x_{mi}y_i \tag{3.5}$$

Summarizing Eqs. (3.3) to (3.5) and expressing in a vector–matrix format, we get

$$\mathbf{X}^T\mathbf{X}\,\vec{b} = \mathbf{X}\,\vec{Y} \tag{3.6}$$

Where, $\mathbf{X} = \begin{bmatrix} 1 & 1 & \cdots & 1 \\ x_{11} & x_{12} & \cdots & x_{1n} \\ \vdots & \vdots & \ddots & \vdots \\ x_{m1} & x_{m2} & \cdots & x_{mn} \end{bmatrix}, \vec{Y} = \begin{bmatrix} y_1 \\ y_2 \\ \vdots \\ y_n \end{bmatrix}, \vec{b} = \begin{bmatrix} b_0 \\ b_1 \\ \vdots \\ b_m \end{bmatrix}.$

In the previous chapter, vectors were represented by non-capitalized bold letters but here, they are capped with a "→" sign.

The reader should expand Eq. (3.6) and confirm that they agree with Eqs. (3.3) to (3.5). Equation (3.6) is known as the characteristic equation of the least squares method. Solving Eq. (3.6) for $\vec{b}$ gives

$$\vec{b} = \left[\mathbf{X}^T\mathbf{X}\right]^{-1}\mathbf{X}\,\vec{Y}. \tag{3.7}$$

Thus we have shown that $[b_0, b_1, \cdots b_m]$ is the solution of Eq. (3.7).

Figure 3.2 illustrates the relationship between the flat surface described by Eq. (3.1) and the plotted dataset when $m = 2$. In this case, the regression equation is that of a flat surface, and the error E defined in Eq. (3.2) can be understood as the distance between the flat surface and the data points, projected onto the y-axis.

3.2 Least Square Solution

Now suppose that n sets of linear equations exist, each with m unknowns, as shown below. To solve this set of simultaneous equations, we must obtain the unknown numbers $X_1, \ldots, X_m$.

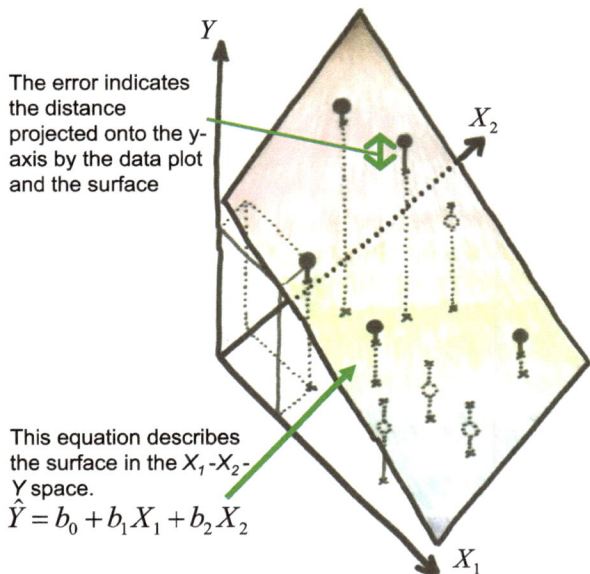

Fig. 3.2 Geometric interpretation of multi regression analysis

The error indicates the distance projected onto the y-axis by the data plot and the surface

This equation describes the surface in the X_1-X_2-Y space.
$$\hat{Y} = b_0 + b_1 X_1 + b_2 X_2$$

$$a_{11}X_1 + a_{12}X_2 + \cdots + a_{1m}X_m = b_1 \qquad (3.8.1)$$

$$a_{21}X_1 + a_{22}X_2 + \cdots + a_{2m}X_m = b_2 \qquad (3.8.2)$$

$$\vdots$$

$$a_{n1}X_1 + a_{n2}X_2 + \cdots + a_{nm}X_m = b_n \qquad (3.8.3)$$

Consistent   with   Eq.   (3.6),   we   adopt   the   following   notation:
$$\mathbf{A} \equiv \begin{bmatrix} a_{11} & \cdots & a_{1m} \\ \vdots & \ddots & \vdots \\ a_{nm} & \cdots & a_{nm} \end{bmatrix}, \vec{X} \equiv \begin{bmatrix} X_1 \\ \vdots \\ X_m \end{bmatrix}, \vec{b} \equiv \begin{bmatrix} b_1 \\ \vdots \\ b_n \end{bmatrix}.$$

Here, if $m > n$, i.e., if the number of unknowns exceeds the number of equations, the system permits an infinite number of solutions, and is hence referred to as indeterminate.

As should be familiar from junior-school algebra, a unique solution is obtained only when $m = n$. Solutions to Eq. (3.8) are then obtained as

$$\vec{b} = \mathbf{A}\vec{X} \Leftrightarrow \vec{X} = \mathbf{A}^{-1}\vec{b}. \qquad (3.9)$$

How do we tackle the situation in which more equations exist than unknowns i.e., $m < n$? We adopt the least squares method, expressing Eq. (3.8) as (the reader should confirm this carefully)

$$^{*}\mathbf{A}\vec{b} = [^{*}\mathbf{A}\mathbf{A}]\vec{X}. \qquad (3.10)$$

Fig. 3.3 Solution of
simultaneous equation with
$m = n = 2$

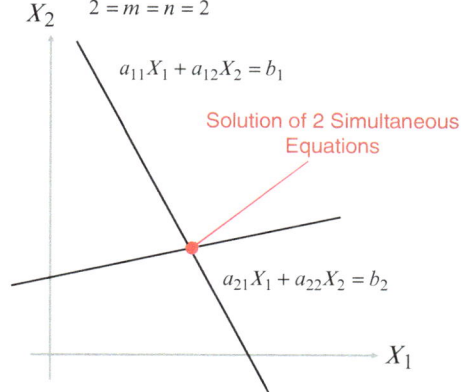

Fig. 3.4 Solution of
simultaneous equation with
$2 = m < n = 5$

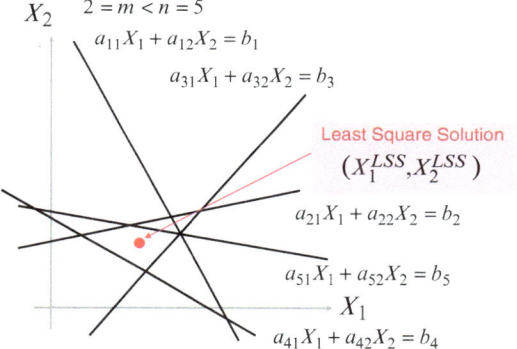

*A is called the **adjoint matrix** of **A**. The adjoint is the transpose of the conjugate complex of **A**. For a matrix of real numbers; the adjoint and transpose are identical. Equation (3.10) is easily solved as

$$\vec{X} = \left[{}^{*}\mathbf{A}\mathbf{A}\right]^{-1}{}^{*}\mathbf{A}\,\vec{b}\,. \tag{3.11}$$

This is the *least square solution*. The matrix $[{}^{*}\mathbf{A}\mathbf{A}]^{-1}{}^{*}\mathbf{A}$, called the *generalized inverse matrix*, reduces to the standard inverse when $m = n$.

Let us present a geometric meaning of least squares fitting. As an example, consider two simultaneous equations involving two unknowns (i.e., $m = n = 2$). Geometrically, the solution is the intersection of the two straight lines described by the assigned equations, as shown in Fig. 3.3. This representation should be familiar to the reader from junior-school algebra.

Now consider the least squares solution when $2 = m < n = 5$, as shown in Fig. 3.4. In this case, we have more equations than unknowns; i.e., excess information is available. As can be understood from the figure, the least squares solution is decided by taking the distance from each perpendicular intersection and

Fig. 3.5 Geometric explanation of least squares solution

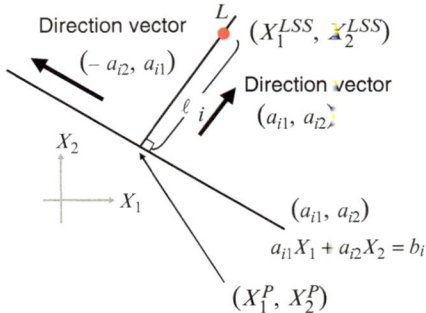

minimizing the sum (more accurately, the sum of the distance squared). This decision is an intuitively agreeable compromise. To confirm that the least squares coordinates equal the sum of squares of the distance from the straight line, consider Fig. 3.5. The best-fit line to the provided dataset is

$$a_{i1}X_1 + a_{i2}X_2 = b_i. \tag{3.12}$$

The equation of the straight line perpendicular to (3.12) which passes through the least squares solution (X_1^{LSS}, X_2^{LSS}), with careful choice of direction vector, is expressed as

$$-a_{i2}X_1 + a_{i1}X_2 = -a_{i2}X_1^{LSS} + a_{i1}X_2^{LSS}. \tag{3.13}$$

Since the base of the perpendicular line (X_1^P, X_2^P) is the intersection between (3.12) and (3.13), Eqs. (3.12) and (3.13) can be solved simultaneously,

$$
\begin{aligned}
X_1^P &= \frac{a_{i2}^2 X_1^{LSS} - a_{i1}a_{i2}X_2^{LSS} + a_{i1}b_i}{a_{i1}^2 + a_{i2}^2}, \\
X_2^P &= \frac{-a_{i1}a_{i2}X_1^{LSS} + a_{i1}^2 X_2^{LSS} + a_{i2}b_i}{a_{i1}^2 + a_{i2}^2}.
\end{aligned} \tag{3.14}
$$

Hence, the length of the perpendicular line is

$$
\begin{aligned}
\ell_i^2 &= \left(X_1^{LSS} - X_1^P\right)^2 + \left(X_2^{LSS} - X_2^P\right)^2 \\
&= \left[\frac{a_{i1}^2 X_1^{LSS} + a_{i1}a_{i2}X_2^{LSS} - a_{i1}b_i}{a_{i1}^2 + a_{i2}^2}\right]^2 + \left[\frac{a_{i1}a_{i2}X_1^{LSS} + a_{i2}^2 X_2^{LSS} - a_{i2}b_i}{a_{i1}^2 + a_{i2}^2}\right]^2 \\
&= \frac{a_{i1}^2\left(a_{i1}X_1^{LSS} + a_{i2}X_2^{LSS} - b_i\right)^2 + a_{i2}^2\left(a_{i1}X_1^{LSS} + a_{i2}X_2^{LSS} - b_i\right)^2}{[a_{i1}^2 + a_{i2}^2]^2} \\
&= \frac{\left(a_{i1}X_1^{LSS} + a_{i2}X_2^{LSS} - b_i\right)^2}{a_{i1}^2 + a_{i2}^2}.
\end{aligned}
$$

$$\tag{3.15}$$

Fig. 3.6 Difference in
geometric interpretation
between multi regression
analysis and the least
squares solution

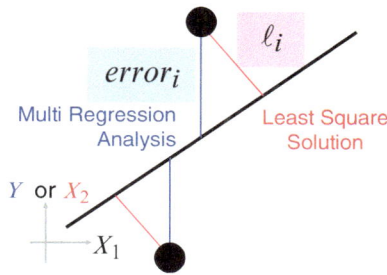

Therefore, the sum of squared errors is:

$$S = \sum_{i=1}^{5} \ell_i^2 = \sum_{i=1}^{5} \frac{\left(a_{i1}X_1^{LSS} + a_{i2}X_2^{LSS} - b_i\right)^2}{a_{i1}^2 + a_{i2}^2}. \tag{3.16}$$

In the general case with m unknowns and n equations, the sum of squared errors
becomes

$$S = \sum_{i=1}^{n} \frac{\left(\sum_{j=1}^{m} a_{ij}X_j^{LSS} - b_i\right)^2}{\sum_{j=1}^{m} a_{ij}^2} \tag{3.17}$$

Similarly to Eqs. (3.3)–(3.5) in the multi regression analysis, we set $\frac{\partial S}{\partial X_j^{LSS}} = 0$
(where j is an integer such that $j \leq m$) to obtain Eqs. (3.10) and (3.11). Readers are
encouraged to derive this result by themselves.

Figure 3.6 illustrates the difference between the geometric interpretation of multi-
regression analysis explained in Fig. 3.2 and that of least square solution introduced
here. We consider two-dimensional data; in other words, two predictor variables in
the multi-regression analysis and two unknown numbers in the least squares solution.
The plot indicates the given dataset for the multi regression analysis and the coordinates of the least square solution. The error in the former is the distance between the
datum and the regression line (projected onto the y-axis), while the latter error is the
geometric distance between the datum and the least-square fitting.

As explained above, multi-regression analysis and least squares solutions principally share a common basis, and the reader has probably understood that both
concepts are based on the least squares method. Moreover, in both methods, the
characteristic equations can be written as vector–matrix equations with the same
structure.

Multi regression analysis and least squares fitting involve inverse matrices,
which are easily obtained using the subroutine MATINV in the program of Sect.
2.13. Ideally, the reader should attempt a general program for multi regression
analysis. In the subroutine MATINV presented in Chap. 2, all real variables are

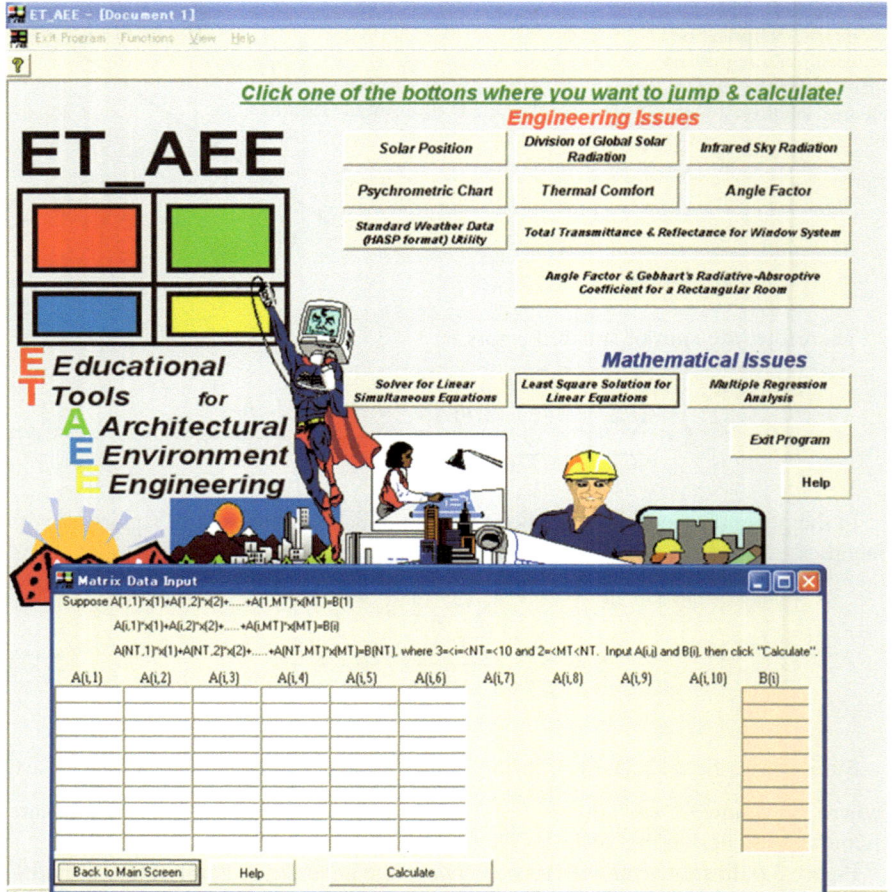

Fig. 3.7 Dialogue screen for least squares solution of ET_AEE

defined to be single-precision; here, double precision is required. The author has developed an environmental engineering calculation tool ET_AEE incorporating the completed general program into an application, whose dialogue screen is shown in Fig. 3.7. The tool can be accessed by downloading the installation file from our research website (http://ktlabo.cm.kyushu-u.ac.jp/).

3.3 Application of Least Squares Solution

This section introduces two applications of least squares fitting. Although the content of this section is not essential, it may be of more interest than that of the previous section.

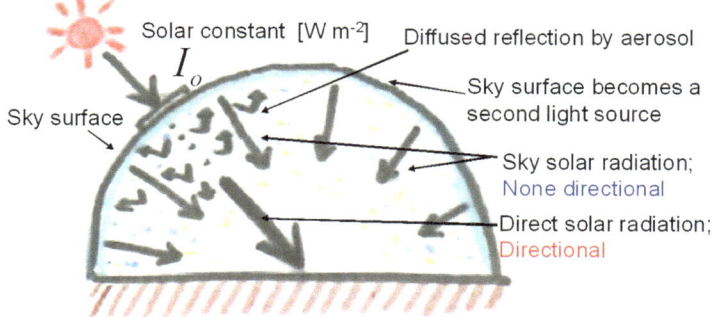

Fig. 3.8   Direct and diffuse components of solar radiation

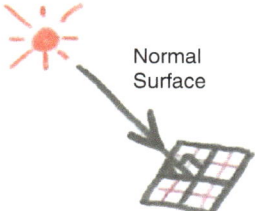

Fig. 3.9   Solar radiation on normal surface

First, we describe the problem of *direct diffuse separation*, in which global solar radiation data are separated into direct and diffuse components. Solar radiation constitutes the electromagnetic waves radiated by the sun, classified as ultraviolet, visible, and infrared radiation. As shown in Fig. 3.8, the direct component of solar radiation maintains its direction when passing through the atmosphere, while the diffuse radiation brightens the entire sky by reflecting from aerosols such as airborne particles and atmospheric vapor. Diffuse radiation reaches the ground using the sky as a secondary light source. On cloudy days, the number of aerosols increases, increasing the proportion of diffuse sky radiation; in contrast, sunny days admit a higher proportion of direct radiation. The solar radiation flux at the surface normal to the boundary of the earth's atmosphere (see Fig. 3.9) is called *the solar constant I_o* [W/m^2]. The value of I_o changes slightly with season but is essentially constant.

Since solar radiation provides large heat flux boundary conditions in all types of environmental systems, predictions must be based on the solar radiation incident on an arbitrary inclined plane. The amount of solar radiation intercepting an inclined plane with *wall surface azimuth α* [deg] and *wall surface angle of inclination θ* [deg] defines the direct component I_D [W/m^2] and diffuse component I_S [W/m^2], respectively (see Fig. 3.10):

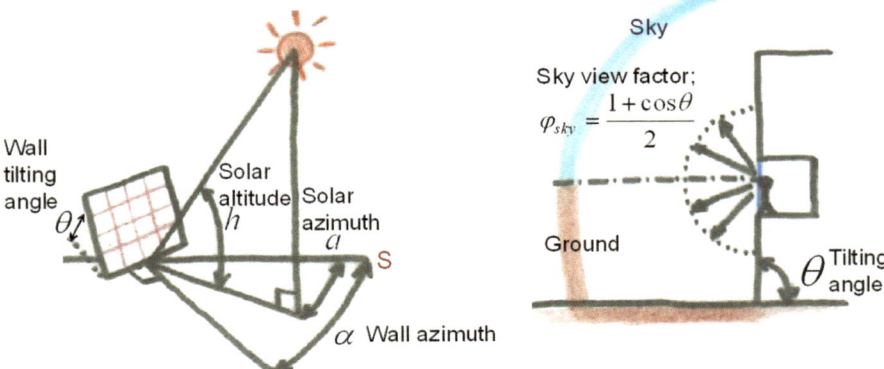

Fig. 3.10 Relationship between arbitrary inclined plane and position of the sun (*left panel*) and the sky factor (*right panel*)

$$\begin{cases} I_D = I_{DN} \cdot \cos i \\ I_S = I_{SH} \cdot \varphi_{sky} = I_{SH} \dfrac{1 + \cos \theta}{2} \end{cases} \tag{3.18}$$

Where, I_{DN} is the *normal surface direct irradiance* [W/m^2], I_{SH} is the *horizontal surface diffuse irradiance* [W/m^2], φ_{sky} is the sky view factor [ND] and i is the *incidental angle to the wall* of the direct solar radiation [deg]. The *sky view factor*, as shown in the right panel of Fig. 3.10, indicates the form factor (a point to a face form factor) viewing the sky from the central (wall) surface; $\varphi_{sky} = 1/2$ at $\theta = 90°$. The angle of incidence can be expressed in terms of the *solar altitude and solar azimuth* (collectively called the *solar position*) and the wall azimuth, as shown in the left panel of Fig. 3.10:

$$\cos i = \cos \theta \sin h + \sin \theta \cos h \cos (a - \alpha). \tag{3.19}$$

The position of the sun is:

$$\begin{cases} \sin h = \sin \phi \sin \delta + \cos \phi \cos \delta \cos t, \\ \sin a = \dfrac{\cos \delta \sin t}{\cos h}, \\ \cos a = \dfrac{\sin h \sin \phi - \sin \delta}{\cos h \cos \phi}, \end{cases} \tag{3.20}$$

where t is the *hour angle*, determined from the present time at the point of calculation (standard time), the longitude of the point where standard time is defined (in Japan, at Akashi-shi, Hyogo Prefecture), the longitude at the point of

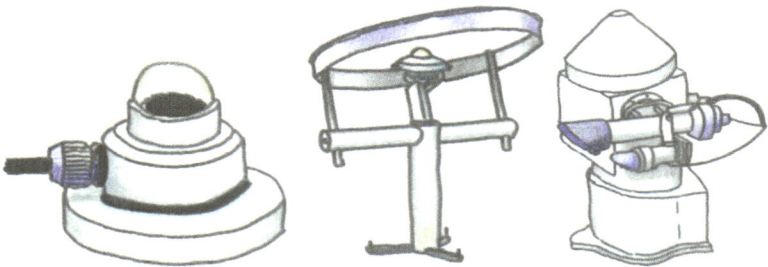

Fig. 3.11 (**a**) Pyrheliometers (*left panel*) (**b**) diffuse irradiance meter with shading ring (*central panel*), (**c**) direct irradiance meter with sun tracking device (*right panel*)

calculation and the *equation of time*, i.e., the difference between the mean and true solar times. The equation of time corrects for the change in angular velocity with season due to Earth's elliptic orbit. δ is the *declination of the sun*, defined as the angle between the Earth's equatorial plane and its plane of revolution. All angles are measured in degrees. δ changes with season, reaching an approximate maximum and minimum of 23.5° and −23.5° at the summer and winter solstices, respectively, and 0° at the vernal and autumn equinoxes. As outlined above,[1] the position of the sun can be determined once the geographic coordinates of the calculation point (latitude, longitude) and date/time are known. In other words, the intensity of solar radiation incident to an arbitrary inclined surface in Eq. (3.18) can be calculated from the normal direct irradiance and the horizontal surface diffuse irradiance. That is, to evaluate the intensity of solar radiation incident on an arbitrary inclined surface, the total solar radiation must be separated into its direct and diffuse components.

Unfortunately, the majority of solar radiation data collected from meteorological offices include only the global horizontal surface solar radiation, from which the direct and diffuse components are difficult to separate. Unmanned observation data collected by AMeDAS and used in weather forecasts lack even the total solar radiation (hours of sunlight alone are recorded). Direct and diffuse solar radiation can be detected simultaneously by placing a pyrheliometer on a horizontal surface, as shown in Fig. 3.11a. From Eq. (3.18), the global horizontal surface solar radiation I_H [W/m^2] is

$$I_H = I_{DN} \cdot \sin h + I_{SH}. \tag{3.21}$$

To separate the diffuse component, in addition to the horizontal surface global irradiance measurements, the horizontal surface diffuse irradiance must be measured by a device equipped with a shading ring, as shown in Fig. 3.11b.

---

[1] For an explanation on how the intensity of solar radiation incident on an arbitrary inclined surface relates to solar position, please read a standard construction environmental engineering or building physics texts.

Fig. 3.12 Bouguer's
equation

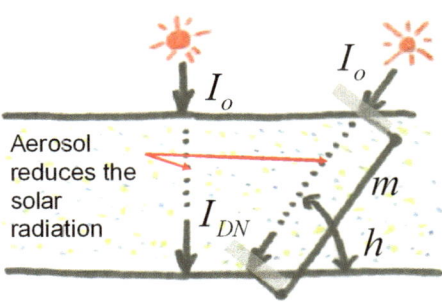

Alternatively, the normal surface direct irradiance is measured using a meter equipped with a sun tracking device (Fig. 3.11c). These specialized measuring devices must also be maintained throughout the measurement period; hence, much care is required in their use. On the other hand, as already mentioned, AMeDAS records hours of sunlight. Previous research has shown that horizontal surface global irradiance can be estimated from these data. If the diffuse and direct components can indeed be extracted from the horizontal surface global irradiance, the intensity of incident solar radiation striking an arbitrarily inclined surface at each AMeDAS point could be calculated, which would be extremely convenient. As evident from Eq. (3.21), this separation problem involves one equation in two unknowns, thus requiring innovative solution approaches. First, we note that the intensity of normal radiation obeys *Bouguer's equation*:

$$I_{DN} = I_o \cdot P^m = I_o \cdot P^{\sin h}. \tag{3.22}$$

As shown in Fig. 3.12, Bouguer's equation describes the theoretical decay of solar radiation propagating through a medium. As already discussed, the solar constant I_0 is known. The *atmospheric permeability P* [ND] indicates the clarity of the atmosphere (with "1" denoting a perfectly clear atmosphere).

To complete the analysis, the horizontal surface diffuse irradiance must be expressed in terms of known parameters (such as atmospheric permeability). Since no theoretical description of diffuse irradiance is known, this quantity must be estimated from semi-empirical or empirical formulae. To this end, we adopt the basic *Berlage formula*:

$$I_{SH} = \frac{1}{2}I_o \cdot \sin h \frac{1 - P^{1/\sin h}}{1 - 1.4\ln P}. \tag{3.23}$$

Equation (3.23) relates the diffuse irradiance to the atmospheric permeability, solar elevation, and solar constant, all of which are known or obtainable. The atmospheric permeability is obtained by substituting Eqs. (3.22) and (3.23) into Eq. (3.21). And by returning to Eqs. (3.22) and (3.23), the diffuse (horizontal surface) and direct (normal surface) irradiance are determined. The Berlage formula assumes that diffuse reflection renders the sky uniformly bright.

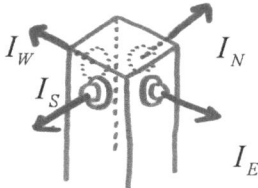

Fig. 3.13 Horizontal surface diffuse irradiance and normal surface direct irradiance can be identified by applying least squares fitting to measured global irradiance data in four perpendicular directions

Consequently, the formula yields reasonable results on sunny days, but introduces a large error on cloudy days. The Berlage formula has been largely replaced by models which reduce the statistical error on cloudy days.[2]

In the above problem, the observed data were the horizontal surface global irradiance data. The direct and diffuse components were separated by assuming a model of diffuse irradiance. Errors in this model will introduce errors in the separation of the components, highlighting the need for accurate and easy separation of the two components from field experiments or field observations. One means of achieving this is least squares fitting. The unknowns are the (normal surface) direct irradiance and (horizontal surface) diffuse irradiance. To increase the number of measured data, the global irradiance is measured on surfaces other than the horizontal surface. For example, suppose that vertical plane global irradiance is simultaneously measured in four directions as shown in Fig. 3.13. When undertaking these measurements, the researcher must ensure that no ground-surface illumination is reflected.

In the vertical plane, Eq. (3.19) becomes

$$\cos i = \cos h \cos (a - \alpha)$$

Hence, if the measured data in the four vertical directions are denoted I_N, I_E, I_S and I_W (note that strict north/south and east/west directions are not required, provided that the wall surface angle is accurately captured), we obtain the following system of simultaneous equations:

$$\begin{cases} I_N = I_{DN} \cdot \cos i_N + 0.5 I_{SH}. \\ I_E = I_{DN} \cdot \cos i_E + 0.5 I_{SH}. \\ I_S = I_{DN} \cdot \cos i_S + 0.5 I_{SH}. \\ I_W = I_{DN} \cdot \cos i_W + 0.5 I_{SH}. \end{cases} \tag{3.24}$$

---

[2] However, the Berlage formula, which remains fundamental to all later models, is sufficient to explain the principles of solar radiation component separation.

Fig. 3.14 Expressing deformed target heat system using single room model

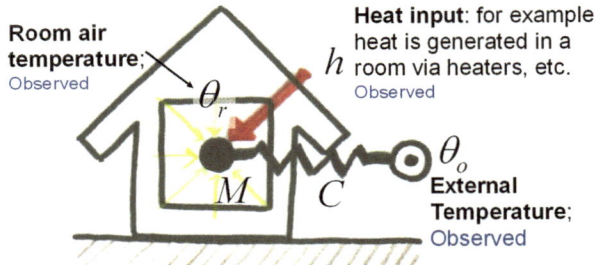

Defining $\mathbf{A} \equiv \begin{bmatrix} \cos\theta_N & 0.5 \\ \cos\theta_E & 0.5 \\ \cos\theta_S & 0.5 \\ \cos\theta_W & 0.5 \end{bmatrix}$, $\vec{X} \equiv \begin{bmatrix} I_{DN} \\ I_{SH} \end{bmatrix}$ and $\vec{b} \equiv \begin{bmatrix} I_N \\ I_E \\ I_S \\ I_W \end{bmatrix}$, the least squares

solution is obtained by solving Eq. (3.11), allowing accurate separation of the direct and diffuse components of solar radiation.

The second example is a *parameter identification problem*, in which the thermal characteristics of existing buildings are estimated from measured data. The target is modeled as a linear heat system involving several undecided parameters. We assume an appropriate physical model whose parameters match the thermal characteristics of the target. The undecided parameters are determined from least squares fitting to the actual data. As the reader will recall from Chap. 2, important thermal characteristics are defined in terms of thermal density and insulation. An example of the latter is the heat loss coefficient [W/(m^{2}K)], which evaluates the loss of heat due to ventilation or leakage through walls. If detailed specifications such as the construction and wall structure of the building can be obtained from plans and other documents, the thermal characteristics are readily obtained. However, for buildings such as old dwellings or storehouses, whose documentation may be missing or lost, alternative evaluation is required.

To simplify the discussion, the target is expressed in a deformed single room model as shown in Fig. 3.14. The heat capacity of the whole building M [J/K] is that at the node where the room temperature has been one-point lumped parameterized. Thermal exchange with external air (such as ventilation and airflow) is expressed as total conductance C [W/K]. In practise, this type of modelling applies only to buildings of small to moderate heat capacity; furthermore, real buildings are not adequately modelled as a single room. Constructing such a deformed physical model of a real building will incur errors that reflect this inadequacy.

Heat flow in the single room model (denoting room and external temperatures by θ_r and θ_o, respectively) is described by

$$M\frac{d\theta_r}{dt} = C(\theta_o - \theta_r) + h, \qquad (3.25)$$

where, h [W] is the heat input into the system. Suppose that measurements are recorded in the target room at time interval Δt. If Eq. (3.25) is time discretized by the Crank–Nicolson method, the following is obtained:

$$M\left[\frac{1}{\Delta t}\left(\theta_r^{j+1} - \theta_r^j\right)\right] = C\left[\frac{1}{2}\left(\theta_o^{j+1} + \theta_o^j\right) - \frac{1}{2}\left(\theta_r^{j+1} + \theta_r^j\right)\right] + \frac{1}{2}\left(h^{j+1} + h^j\right)$$

$$\Leftrightarrow \left[\frac{1}{\Delta t}\left(\theta_r^{j+1} - \theta_r^j\right)\right]M + \left[-\frac{1}{2}\left(\theta_o^{j+1} + \theta_o^j\right) + \frac{1}{2}\left(\theta_r^{j+1} + \theta_r^j\right)\right]C = \frac{1}{2}\left(h^{j+1} + h^j\right).$$

$$(3.26)$$

Both terms inside the large bracket on the left-hand side involved measured data. The terms on the right-hand side of (3.26) can also be measured. Therefore, if numerous measured datasets are obtained, the unknowns M and C can be derived from least squares data fitting. In vector–matrix notation, Eq. (3.26) is expressed as (3.27), from which $\begin{bmatrix} M \\ C \end{bmatrix}$ can be solved from Eq. (3.11).

$$(3.27)$$

To correctly apply the above method, measurements should be performed during the night, when solar radiation is absent. Furthermore, one must account for the heat h supplied by the heater. As shown in Fig. 3.15, in the absence of the heater, the measured data will depend strongly on conditions, thereby destabilizing the identified least squares solution. If transient data are collected when the heater is switched from off to on (and vice versa) a stable solution should arise. This scenario reflects the so-called multi-collinearity problem in multi-regression analysis; if the statistical model includes predictor variables with high dependency, the regression becomes unstable.

Suppose that heat characteristic parameters are identified for a large, diverse sample set. If the obtained data are plotted, what do we observe? Buildings such as traditional houses or storehouses have large heat capacity with poor insulating ability, whereas modern apartments (which are super airtight and highly insulated) should display opposite characteristics. Figure 3.16 is a hypothetical example of the clustered distributions that might arise in this case. The interested reader could peruse such clustering as a research topic.

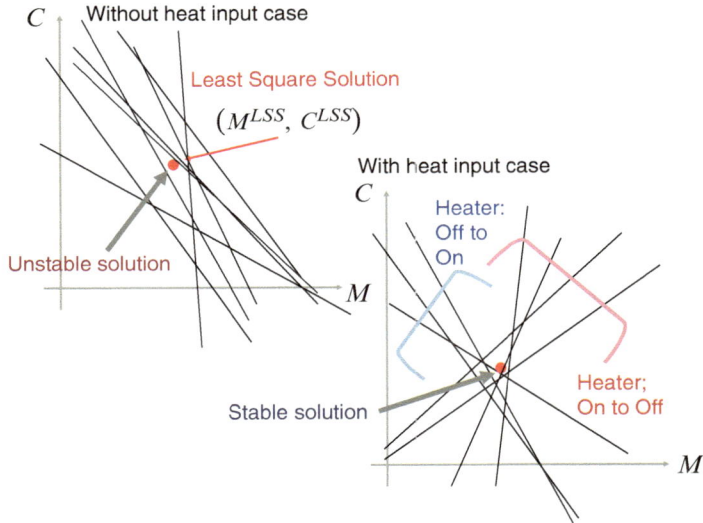

Fig. 3.15 Dataset with strong dependency (*left panel*) and dataset with established independence (*right panel*)

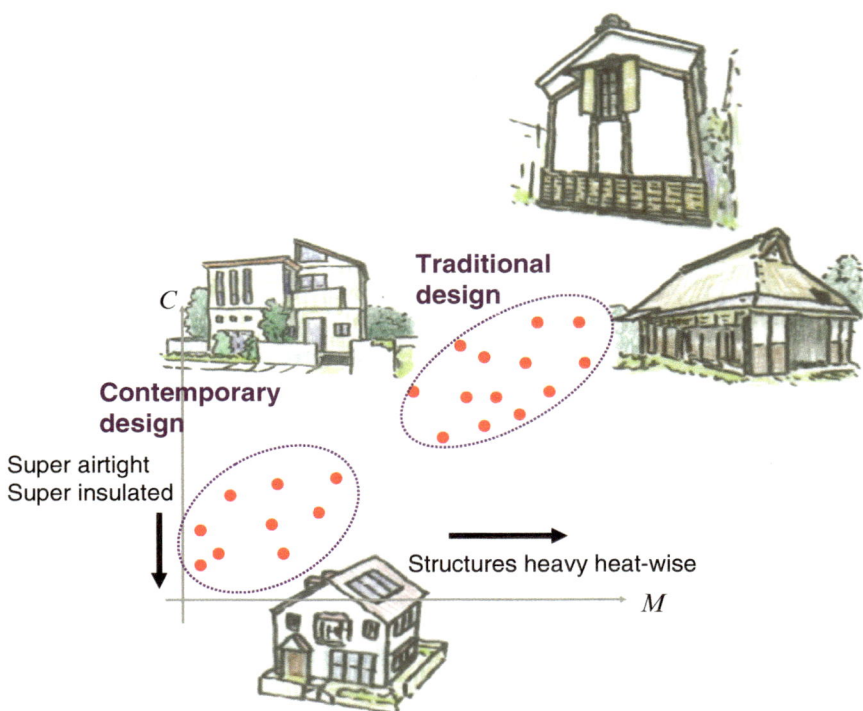

Fig. 3.16 If M and C with identified parameters are plotted in various situations

Chapter 4
Dynamics of Non-Linear Systems

Abstract In Chap. 2, we introduced linear system state equations. In this chapter, the discussion is extended to non-linear systems and their general dynamic properties. While Chap. 2 primarily adopted a numerical approach, here we focus on the deductive approach. In the latter half of the chapter, evolutionary games are introduced as a template for discussion.

Keywords 2-Palyer 2-strategy evolutionary game • Non-linear dynamical systems

4.1 Linear Dynamical Systems

Let us revisit the system state Eq. (2.18). Similar to the numerical stability analysis of Sect. 2.6, the boundary conditions are not considered. As already explained, a boundary condition operates externally to the system (in this case, via a "temperature raising" mechanism) and is not related to the intrinsic dynamics of the system.

$$\frac{d\mathbf{x}}{dt} = \dot{\mathbf{x}} = \mathbf{Ax}. \tag{4.1}$$

Equation (4.1) is in a linear format. By linear format[1] we mean that the time evolution of the system is described by a vector matrix operation. In other words, in a linear system, the elapsed time in the system (dynamics) can be described by the familiar linear algebra introduced at senior school.

What happens to $\mathbf{x}$ in Eq. (4.1) as $t \to \infty$? One might imagine that changes will occur until $\frac{d\mathbf{x}}{dt} = \mathbf{0} \Leftrightarrow \dot{\mathbf{x}} = \mathbf{0}$, denoting a state of no further change. This eventual state, called steady state in many engineering fields, is called *equilibrium* in physical

---

[1] The "linear" quality of a system is truly beneficial in engineering. No sharp fluctuations develop over time; therefore, future behavior is easily extrapolated from currently available information.

dynamical systems (or in fields such as economics). Hence, the equilibrium state is defined as $\dot{\mathbf{x}} = \mathbf{0}$. The equilibrium point is frequently expressed as $\mathbf{x}^*$.

By treating Eq. (4.1) as an ordinary scalar differential equation, its solutions are obtained as

$$\frac{d\mathbf{x}}{dt} = \mathbf{A}\,\mathbf{x} \Leftrightarrow \frac{1}{\mathbf{x}}d\mathbf{x} = \mathbf{A}dt \Leftrightarrow \mathbf{x} = \exp[\mathbf{A}t] + \mathbf{c}, \qquad (4.2)$$

where $\mathbf{c}$ is an integration constant vector. At equilibrium, $\dot{\mathbf{x}} = \mathbf{0} \Rightarrow \mathbf{A}\mathbf{x}^* = \mathbf{0} \Leftrightarrow \mathbf{x}^* = \mathbf{0}$. Under what circumstances will $\mathbf{x} \to \mathbf{0}$ as $t \to \infty$ in Eq. (4.2)? Let us once again use the analogy with scalar cases. Evidently, the solutions $x(t) = \exp[at] \to 0$ as $t \to \infty$ if and only if $a < 0$. Vector matrix systems of equations are solved similarly, by finding the eigenvalues of the matrix $\mathbf{A}$. We have already discussed similar logic in the numerical stability analysis of Sect. 2.6. If the equilibrium point in Eq. (4.1) is to satisfy $\mathbf{x} \to \mathbf{0}$, all n eigenvalues of the $n \times n$ matrix $\mathbf{A}$ must be negative. Thus, to explain the equilibrium situation in Eq. (4.1), we should examine each eigenvalue in the transition matrix $\mathbf{A}$, which determines the time evolution of the system.

To simplify the discussion without loss of generality, we suppose that $\mathbf{A}$ is a 2×2 matrix with eigenvalues λ_1 and λ_2. Three sign combinations of these eigenvalues are possible; both positive, both negative, or one positive and one negative. The signs of the eigenvalues determine the stability of the equilibrium point $\mathbf{x}^* = \mathbf{0}$ in our current problem, as illustrated in Fig. 4.1. When all eigenvalues are negative, the equilibrium point $\mathbf{x}^*$ is stable(in Eq. (4.1), $\mathbf{x}^* = \mathbf{0}$). In stable equilibrium, $\mathbf{x}^*$ behaves like a jug whose potential is minimized at its base, so that all points surrounding $\mathbf{x}^*$ are drawn toward it. In Eq. (4.1), with a single equilibrium point at $\mathbf{x}^* = \mathbf{0}$, the system eventually converges to $\mathbf{x}^* = \mathbf{0}$ regardless of the initial conditions. If all eigenvalues are positive then $\mathbf{x}^* = \mathbf{0}$ behaves like the peak of a dune (see central panel of Fig. 4.1). In this case, regardless of the initial conditions, the system never attains $\mathbf{x}^* = \mathbf{0}$, and the system is unstable. If both positive and negative eigenvalues exist, $\mathbf{x}^* = \mathbf{0}$ converges in one direction but diverges in a linearly independent direction, as shown in the right panel of Fig. 4.1. Such an equilibrium point is called a *saddle point* (viewed three-dimensionally in Fig. 4.2), and is also unstable.

In summary, the equilibrium point is the solution of the given system state equation satisfying $\dot{\mathbf{x}} = \mathbf{0}$. The signs of the eigenvalues of the transition matrix determine whether the equilibrium point $\mathbf{x} = \mathbf{x}^*$ is a source, a sink, or a saddle point. Negative and positive eigenvalues give rise to sinks and sources, respectively, while mixed eigenvalues signify a saddle point. This seemingly trivial fact is of critical importance. Once the nature of the equilibrium points of a system is determined, laborious numerical calculations to find stationary solutions are not required. Estimating the system dynamics by closely examining the eigenvalues is known as the deductive approach. To reiterate, if a deductive approach is possible, there is no requirement for numerical solutions.

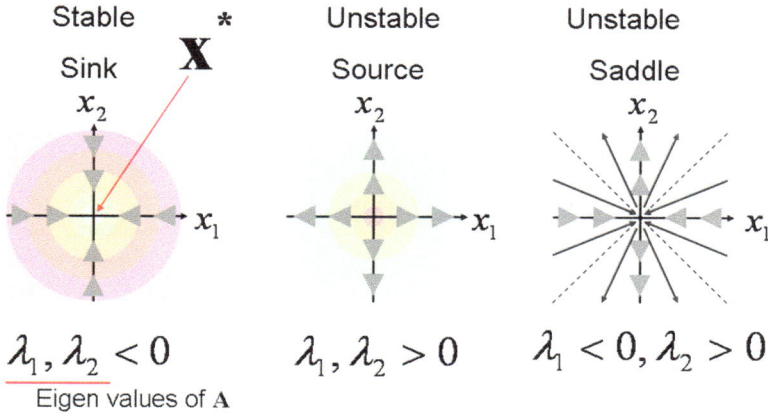

Fig. 4.1 Characteristics of equilibrium point

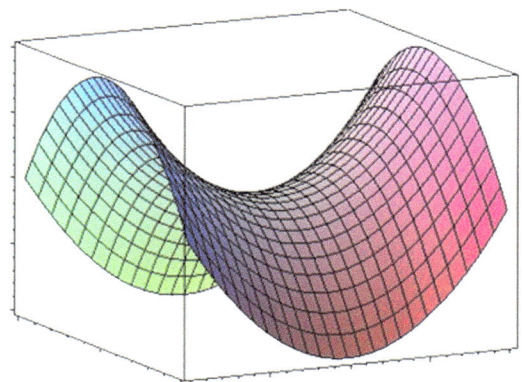

Fig. 4.2 Saddle

Thus far, Eq. (4.1) has been considered as continuous in time. We now reinterpret (4.1) as a time-discretized system and investigate its behavior. The essence of time discretization was explained in Chap. 2.

Initially, we adopt a forward difference scheme in time. Equation (4.1) becomes

$$\mathbf{x_{k+1}} - \mathbf{x_k} = \Delta t \cdot \mathbf{A}\,\mathbf{x_k} \Leftrightarrow \mathbf{x_{k+1}} = (\Delta t \cdot \mathbf{A} + \mathbf{E})\mathbf{x_k}. \tag{4.3}$$

In physical dynamical systems, a recurrence equation such (4.3), in which a linear continuous equation is discretized in time, is sometimes called a *linear mapping*. The transition matrix $\Delta t \cdot \mathbf{A} + \mathbf{E} \equiv \mathbf{T}$ of Eq. (4.3) is essentially equal to Eq. (2.32). For this linear mapping to be stable (non-diverging), the absolute value of the maximum eigenvalue of the transition matrix must not exceed 1. Thus, the necessary and sufficient stability criterion is as follows: (see also Eq. (2.30))

$$|\text{Max}[\text{eigen}[\mathbf{T}]]| \leq 1.$$

Now, let us assume stability as an original system characteristic. In other words, assume that the following is true:

$$\text{Max}[\text{eigen}[\mathbf{A}]] \leq 0. \tag{4.4}$$

The eigenvalue of the unit matrix $\mathbf{E}$ is 1. On page 17, we mentioned that if the eigenvalues λ_D of a matrix $\mathbf{D}$ are known, the eigenvalues of a function of $\mathbf{D}$, $f(\mathbf{D})$, are $f(\lambda_D)$. Applying this rule under the assumptions of Eq. (4.4), the transition matrix of the linear mapping becomes

$$\text{Max}[\text{eigen}[\mathbf{T}]] < -1. \tag{4.5}$$

Equation (4.5) suggests that even when Eq. (4.4) holds, $|\text{Max}[\text{eigen}[\mathbf{T}]]| \leq 1$ is not necessarily satisfied. Thus, the linear mapping of an originally stable system may be unstable. This is a surprising result. It implies that even though the original qualities were good, the calculations fail because of errors introduced in subsequent "time discretization" operations. This potential instability, generated when continuous time is mapped to a discrete system, is exactly the numerical instability detailed in Sect. 2.6. We now consider the same linear mapping under backward difference time discretization. In this case, the mapping is

$$\mathbf{x_{k+1}} - \mathbf{x_k} = \Delta t \cdot \mathbf{A} \, \mathbf{x_{k+1}}$$
$$\Leftrightarrow \mathbf{x_{k+1}} = [1 - \Delta t \cdot \mathbf{A}]^{-1} \mathbf{x_k} = \mathbf{T} \mathbf{x_k} \tag{4.6}$$

from which we obtain

$$0 < \text{Max}[\text{eigen}[\mathbf{T}]] < 1. \tag{4.7}$$

This linear mapping never diverges and will not cause the numerical fluctuations discussed in Sect. 2.7. Thus, if the original qualities are good, it appears that the integrity of the system is retained under backward difference time discretization.

4.2 Non-Linear Dynamical Systems

Consider a continuous dynamical system in which the system state equations are expressed by a non-linear function $\mathbf{f}$:

$$\frac{d\mathbf{x}}{dt} = \dot{\mathbf{x}} = \mathbf{f}(\mathbf{x}). \tag{4.8}$$

The subsequent procedure is typical of how nonlinearities are treated in all types of analyses. Non-linear functions are approximated to linear functions over infinitesimal intervals by Taylor expansion. Expanding the right hand side of Eq. (4.8), we get

$$\mathbf{f}(\mathbf{x}) = \mathbf{f}\,(\mathbf{x}^*) + \mathbf{f}'(\mathbf{x}^*)(\mathbf{x} - \mathbf{x}^*) + \frac{\mathbf{f}''(\mathbf{x}^*)}{2!}(\mathbf{x} - \mathbf{x}^*)^2 + \cdots$$

$$\Leftrightarrow \mathbf{f}(\mathbf{x}) \cong \mathbf{f}\,(\mathbf{x}^*) + \mathbf{f}'(\mathbf{x}^*)(\mathbf{x} - \mathbf{x}^*). \tag{4.9}$$

From the definition of equilibrium point, $\mathbf{f}\,(\mathbf{x}^*) = 0$ (this should be evident by substituting $\frac{d\mathbf{x}}{dt}\big|_{\mathbf{x}=\mathbf{x}^*} = 0$ in Eq. (4.8)), Eq. (4.9) is approximately equal to

$$\mathbf{f}(\mathbf{x}) = \mathbf{f}'(\mathbf{x}^*)(\mathbf{x} - \mathbf{x}^*). \tag{4.10}$$

Equation (4.10) is approximated to a linear equation as follows:

$$\mathbf{f}(\mathbf{x}) = \mathbf{f}'(\mathbf{x}^*)(\mathbf{x} - \mathbf{x}^*) = \mathbf{f}'(\mathbf{x}^*)\mathbf{x} - \mathbf{f}'(\mathbf{x}^*)\mathbf{x}^*. \tag{4.11}$$

The first term on the right of (4.11) is first-order in $\mathbf{x}$, while the second term is constant. Now we can apply the deductive approach introduced in the previous section. Clearly the transition matrix is $\mathbf{f}'(\mathbf{x}^*)$. We must determine the signs of the eigenvalues corresponding to the equilibrium points of this matrix.

The transition matrix is the *Jacobian matrix* of tangent gradients of the multi-variable vector function.

$$\mathbf{f}'(\mathbf{x}^*) = \frac{\partial \mathbf{f}(\mathbf{x})}{\partial \mathbf{x}}\bigg|_{\mathbf{x}=\mathbf{x}^*} = \begin{bmatrix} \frac{\partial f_1(\mathbf{x})}{\partial x_1} & \cdots & \frac{\partial f_1(\mathbf{x})}{\partial x_n} \\ \vdots & \ddots & \vdots \\ \frac{\partial f_n(\mathbf{x})}{\partial x_1} & \cdots & \frac{\partial f_n(\mathbf{x})}{\partial x_n} \end{bmatrix}_{\mathbf{x}=\mathbf{x}^*} \tag{4.12}$$

Let us apply the deductive procedure of Sect. 4.1 to the non-linear system state Eq. (4.8). First, we seek the equilibrium points of Eq. (4.8), which are solutions to $\dot{\mathbf{x}} = \mathbf{0}$ in the given system state equation. A system may contain one or several equilibrium points. In general, quadratic and quartic non-linear functions possess two and four equilibrium points, respectively. Whether each of these equilibrium points ($\mathbf{x} = \mathbf{x}^*$) is a source, a sink, or a saddle point is determined by the sign of the eigenvalues of the transition matrix (4.12). As before, if all n eigenvalues are negative, the equilibrium point is a stable sink, if all are positive, it is an unstable source, and if a mix of signs is found, it is an unstable saddle point. The stability characteristics of the equilibrium points apply only within the vicinity of the equilibrium points (as assumed in the Taylor expansion). Hence, when several equilibrium points exist, the behavior of the system as $t \to \infty$ depends on the starting point of the dynamics, i.e., the initial values. Because the linear system in Sect. 4.1 possessed a single equilibrium point at $\mathbf{x}^* = \mathbf{0}$, this type of initial condition dependency was irrelevant, but non-linear systems can depend heavily on the initial conditions.

4.3   2-Player 2-Strategy Evolutionary Game

In this section, the *2-player 2-strategy game* (abbreviated as *two-by-two game*) is presented as an example of a non-linear system. As the reader will come to appreciate, this apparently esoteric two-by-two game is related to environmental problems.

As previously explained, the two-by-two game is a branch of applied mathematics that models human decision making. It is a relatively new mathematical tool based on the pioneering work of von Neumann and Morgenstern entitled "Theory of games and economic behavior"[2] published in 1944. The applications of the two-by-two game are extremely diverse, ranging from social sciences such as economics and politics to biology, information science, and physics. If a group of particles possessing binary strategies of cooperation or defection is imposed to develop a spatial structure, clusters of cooperation particles emerge abruptly. This seems similar to formation of crystallization or phase transitions in materials. Currently, these analogies have drawn huge interest from members of the statistical physics community.

From an unlimited population, two individuals are selected at random and made to play the game. The game uses two discrete strategies (as shown in Fig. 4.3); cooperation (C) and defection (D). The pair of players receives payoffs in each of the four combinations of C and D. A symmetrical structure between the two players is assumed. In Fig. 4.3, the payoff of player 1 (the "row" player) is represented by the entries preceding the commas; the payoff of player 2 (the "column" player) by the entries after the commas. The payoff matrix is denoted by $\begin{bmatrix} R & S \\ T & P \end{bmatrix}$. A player can also be called an agent. Depending on the relative magnitudes of the matrix elements P, R, S, and T, the game can be divided into 4 classes; the *Trivial* game with no dilemma, the *Prisoner's Dilemma* (sometimes abbreviated to *PD*), *Chicken* (also known as Snow Drift Game or Hawk–Dove Game) and *Shag Hunt* (sometimes abbreviated to SH). The main aim of this section and the next is to show that these

Agent1   Agent2

R; Reward, T; Temptation,
S; Sucker, P; Punishment

Agent2 \ Agent1	Cooperation (C)	Defection (D)
Cooperation (C)	R, R	S, T
Defection (D)	T, S	P, P

Fig. 4.3 Payoff matrix of two by two game

---

[2] Chikuma Scholastic Collection has published a three part Japanese edition of this seminal work (Ginbayashi et al.), reprint (2009).

Game class	Dilemma?	GID	RAD
Prisoner's Dilemma; PD	Yes	Yes	Yes
Chicken (Snow Drift; Hawk-Dove)	Yes	Yes	No
Stag Hunt; SH	Yes	No	Yes
Trivial	No	No	No

Fig. 4.4 Class type in two-by-two game

four game classes can be derived from the eigenvalues of the system per deductible approach for non-linear system equation explained in the previous section.

Here, the gamble-intending dilemma (hereafter referred to as GID) and risk-averting dilemma (hereafter referred to as RAD) are introduced. The existence of these dilemmas is determined by D_g and D_r, defined as follows:

$$D_g \equiv T - R.$$
$$D_r \equiv P - S. \tag{4.13}$$

If $D_g > 0$, GID behavior results, while $D_r > 0$ leads to RAD. Each of the dilemma classes and the existence of GIDs and RADs are summarized in Fig. 4.4. Although, the reader may be overwhelmed at this point having been introduced to a large set of qualities without proofs or detailed explanations, we request the reader to bear with this for just a little bit longer. GIDs are sometimes called Chicken dilemmas while RIDs can be referred to as SH dilemmas. Figure 4.4 shows that the PD game may be Chicken or SH (details will be provided later).

A couple of further explanations are needed here.

Figure 4.5a shows a game setup of the prisoner's dilemma (PD) class. Calculating D_g and D_r from Eq. (4.13), both eigenvalues are seen to be positive; thus, from Fig. 4.4, the game is PD, for reasons which will be explained later. For now, examine panel (b) in Fig. 4.5. The payoff values before the commas, i.e., those of the row-represented agent, are shaded orange and green. In these situations, the column agent is fixed in strategy C or D. The larger of the two elements shaded with the same colour is marked in bold text. These bold values denote whether C or D is the more rational choice for the row agent. Panel (c) illustrates a similar scenario with fixed row agent, indicating whether C or D is the most rational strategy for the column agent. In panel (d), the element for which both row and column agents appears bold is shaded red. The state thus obtained (the game outcome) is known as the *Nash equilibrium*. In this example, the Nash equilibrium indicates the grouping of rational strategies adopted by an agent selected at random from an unlimited group who participates in a single game. Figure 4.5 reveals that both agents exhibit D behavior, and defect each another to accept low profit P (also from that figure, the relationship $T > R > P > S$ is seen to hold in PD). Relating this outcome to the non-linear dynamics of the previous section, even if the unlimited group began with an even division of cooperative and defection agents (50 % cooperators & 50 % defectors), once the game is

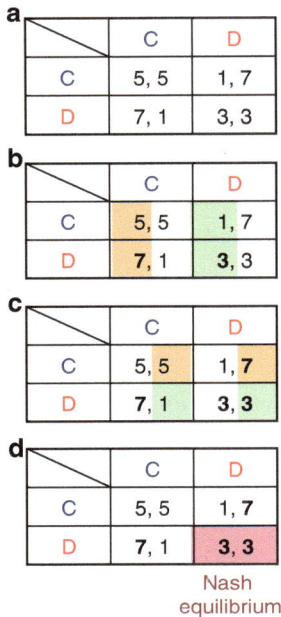

Fig. 4.5 Derivation method for Nash equilibrium with PD as an example

started and the strategy of the agents reviewed according to a certain set of rules after every step[3]; as time progresses,[4] the system will stabilize into a state in which all members (despite the unlimited population size) exhibit defection behavior.

Figure 4.6 plots the payoffs for Agents 1 and 2 on the vertical and horizontal axis, respectively, and displays the payoff matrices for each of the four game classes. These diagrams show the *feasible solutions regions*. The pink areas within the feasible solutions of PD and Chicken reside in the 1st, 2nd, and 4th quadrant (around the central point R). When several plots exist in these regions, we hope to determine the most desirable game outcome between the equal outcomes of Agents 1 and 2. In reality, T and S are clearly the desirable outcomes for Agent 1 and his opponent, respectively. However, we have seen that both agents compromise by taking the fair option R, rather than seeking maximum payoffs for themselves. In this case, R is not the optimal solution but is merely a *fair Pareto optimization*. In contrast to this, in SH and Trivial games, R is the only possible outcome in the pink region (result not shown), and a unique optimal solution exists, R.

In Fig. 4.6, the open and filled circles ○ and ● indicate that Agent 1 (your own hand, say), adopts C and D strategies, respectively. The C and D strategies of Agent 2 (the opponent's hand) are delineated by gray and black dotted lines, respectively.

---

[3] This is referred to as strategy adaptation.

[4] This is referred to as evolution.

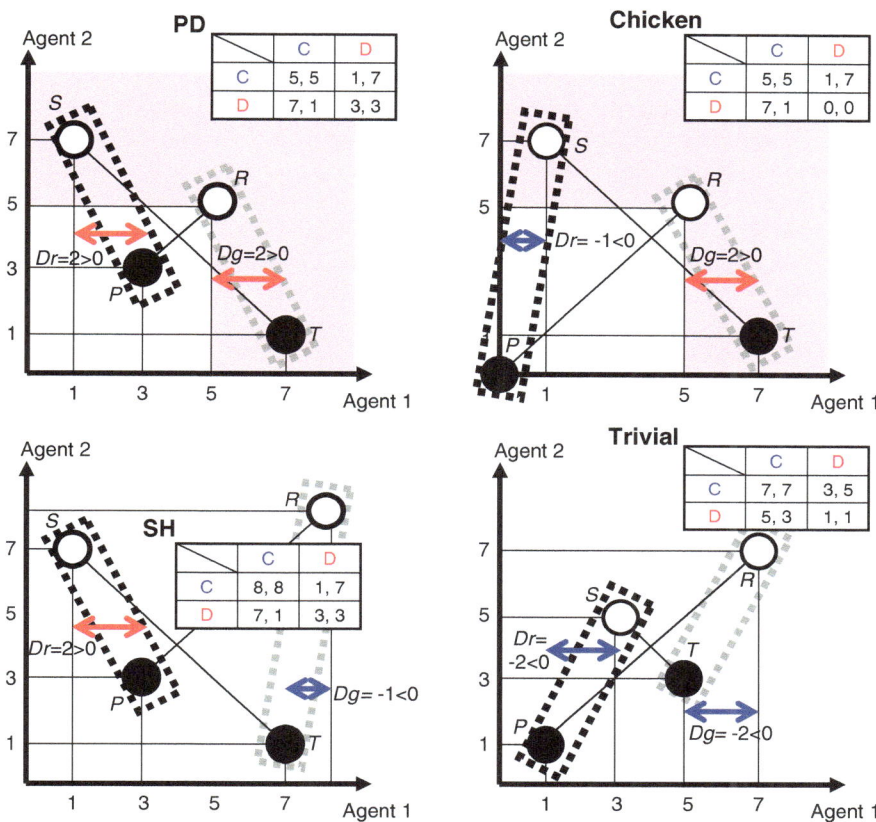

Fig. 4.6 Feasible solution regions of each game class and examples of D_g and D_r

With this visualization, the following discussion should be apparent. In the PD game (upper left panel of Fig. 4.6), if the strategy of the opponent's hand is fixed as C (region within the gray dotted lines), the most rational strategy for your hand is D, which lies further along the horizontal axis (indicating a higher payoff for Agent 1). If your opponent's hand is fixed on D, the same situation arises; within the D region of Agent 2, the D strategy of Agent 1 lies further along the horizontal axis than the C strategy. In other words, you should adopt the D strategy regardless of your opponent's behavior, and the system settles into Nash equilibrium. Similarly for the Trivial game (lower right panel of Fig. 4.6), comparing the areas enclosed by black and grey dotted lines, we observe that Agent 1 should adopt the C strategy regardless of the opponent's hand, and that Nash equilibrium is the R outcome (C, C). The Nash equilibria in the Chicken and SH games are obtained from the payoff matrices as explained in Fig. 4.5. The Nash equilibria in Chicken are the S and T outcomes (C, D) and (D, C), while in SH, they are the R and P outcomes (C, C) and (D, D). In Chicken and SH, the Nash equilibria cannot be determined from the feasible solution regions in Fig. 4.6, but whether one's own strategy should change in response to the opponent's strategy (C or D) can be gauged from the horizontal axis's value of the plots surrounded by black or gray (see upper right and lower left panels of Fig. 4.6 for Chicken and SH games, respectively).

The above dilemmas, to which we have referred so extensively, are defined in the following paragraphs.

A numerical dilemma is introduced whenever the Pareto optimization does not match the Nash equilibria. In PD, Chicken, and SH, the fair Pareto optimizations differ from the Nash equilibria. SH yields only partial match ((C, C) is one of Nash equilibria), but causes dilemma because other outcomes are also possible. The details are explained below.

In PD, the magnitudes of the outcomes are $T > R > P > S$. In reverse phrasing, the order $T > R > P > S$ characterizes the PD game class. Since D_g and D_r are both positive, GIDs and RADs may coexist. The Chicken dilemma, an alternative name for the former, arises from the positive value of $D_g = T - R$. However, as evident from the regions of feasible solutions in the PD and Chicken games shown in Fig. 4.6, when this condition is satisfied, T and S always exist in the first, second, and fourth quadrants (assuming R as the center). Thus, it could be argued that "an incentive to exploit the opponent" exists. In a similar vein, positive $D_r = P - S$ leads to the SH dilemma. However, when this condition is satisfied (results not schematically shown with color highlight), the feasible solution regions in Fig. 4.6 become that T and S always exist in the second, third, and fourth quadrants (assuming P as the center), suggesting "an incentive of not being exploited by the opponent." In fact, this situation emerged in the PD dynamics discussed earlier; as $t \rightarrow \infty$, the entire population became defection. Such an equilibrium state is called *D-dominate*.

In the Chicken game, $T > R > S > P$. Since $D_g > 0$ and $D_r < 0$, the gamble-intending (Chicken-type) dilemma exists in the absence of the risk-averting (SH-type) dilemma. In this game, you incur little risk of being ruined by your opponent but you may gain an advantage by exploiting the opponent. The Chicken game is characterized by $S > P$. That is, the most convenient situation for yourself would arise if you and your opponent adopt the D and C strategies, respectively ($T > R$). Conversely, if you and your opponent both adopt the D strategy, the worst outcome(P, P) results. Being ruined by your opponent would be a more favorable scenario ($S > P$). The structure of environmental issues is very similar. The environment is a public property available to anyone, but if overused by all individuals, it gets depleted. To preserve the environment, individuals might benefit from not using it, and hence a social dilemma is created. This supposed environment may be regarded as a public pastureland, from which your cows may be permitted to consume an unlimited or restricted amount (corresponding to defection and cooperation strategies, respectively). In the short-term, the cooperative strategy restricts the cows' diet until the ground has recovered. This situation can be modelled as a multi-player Chicken game termed the *tragedy of commons* [1]. The Nash equilibria of the Chicken game are (C,D) and (D,C), implying that if half of the population are initially cooperative,[5] as $t \rightarrow \infty$, cooperation and defection members exist in certain proportions (this does not mean that specific agents are restricted to C and D strategies, but rather that the proportions of individuals adopting C and D stabilize to fixed values). This scenario is called *coexistence or polymorphic equilibrium.*

The SH game is characterized by $R > T > P > S$. Since $D_g < 0$ while $D_r > 0$, risk-averting (SH-type) dilemmas exist in the absence of gamble-intending (Chicken-type) dilemmas. Although there is no incentive to exploit one's opponent (since R is optimal and $R > T$), an individual risks damage from an opponent ($P > S$). For instance, if two hunters cooperate to secure a large catch, such as a deer, a successful outcome is likely. However, if the opponent is not certain to cooperate (but instead might defect to cause trouble for the co-operator while knowingly losing their share of the catch), the dilemma of whether one should go on a rabbit hunt (which can be undertaken single-handedly, and is a defection strategy) arises. The name "deer hunting game" is derived from this episode in Chapter Two of "Discourse on Inequality" by Jean-Jacques Rousseau, who is famous for "The Social Contract" and "Émile." The deer hunting game epitomises SH. The Nash equilibria in SH are (C,C) and (D,D), but the dynamics depend on the initial proportion of cooperative individuals. As $t \rightarrow \infty$, the systems converge to either complete defection or complete cooperation. In other words, whether a dark, uncooperative society or a fully cooperative society emerges depends on the initial proportion of cooperators. This type of dynamics is known as *bi-stable.*

In the Trivial game, $R > T > S > P$, and D_g and D_r are both negative. This system is devoid of GIDs and RADs. The Nash equilibrium matches the optimal solution (C, C); thus, regardless of initial cooperation status, all members become cooperative as $t \rightarrow \infty$. This type of equilibrium is called *C-dominate.*

The PD game presents tough dilemmas containing both Chicken and SH-type dilemmas. Since a portion of the optimal SH solutions matches the Nash equilibria, the SH dilemma is weaker than the Chicken dilemma. As previously explained, whether a fully cooperating society emerges depends upon the initial values.

Here a quick diversion will be made.

What governs the evolution of games logic which has aroused such enthusiasm to study by mathematicians, biologists, physicists, information scientists, and even common foot soldiers such as the author? At the end of the day, this comes down to a question: What additional mechanisms will promote ultimate cooperation among the agents if a pair of agents is randomly selected from an unlimited group (an infinite and *well-mixed* selection) and forced into a specified game (such as PD)? In the natural world, cooperative behaviour is found not only in human societies but also among social insects such as ants and bees. This question invokes the mysteries of biological evolution, and invites analogies with statistical physics of crystal structure and phase transitions. Solutions may lead to suggestions for an improved human society.

---

[5] The proportion of cooperative members at the start of a series of games is 0.5.

Fig. 4.7 Five basic
mechanisms of dilemma
resolution and example of
network reciprocity

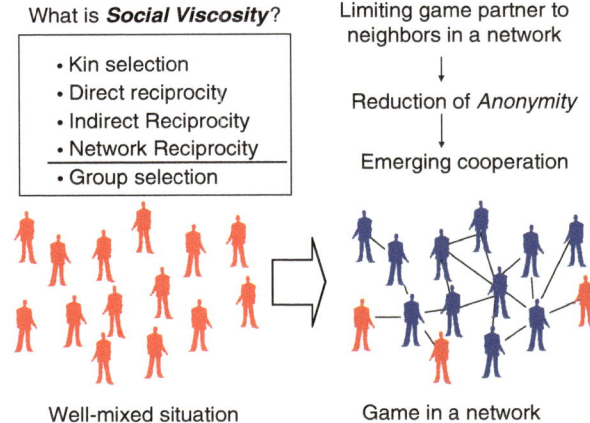

From recent theoretical studies, a rough "supplementary framework" of dilemma resolution can be proposed. Nowak [2] showed there are the five fundamental protocols to mitigate or cancel dilemmas,[6] summarized as in Fig. 4.7. The mechanisms of these activities are governed by very ordinary and beautiful mathematical expressions similar to those of kin selection [3]. Nowak refers to these dynamics as "*Social Viscosity*." Under these circumstances, the population is initially well-mixed as before, and each game is played by a single person whose next encounter is unknown. But, in repeated game battles between a pair of individuals (direct reciprocity),[7] or observing the tag of the opponent (indirect reciprocity), the behaviour of opponent; cooperation or defection, can be distinguished. Or, when players play games against only the neighboring players throughout the network, information relating to strategy is obtained (network reciprocity). All these enable the agents to overcome the dilemmas and create a cooperative society.[8] These processes essentially reduce the anonymity from that of a infinite and well-mixed population (which exists in a total anonymous state) and authenticate the battle opponent. By carefully studying the authentication of others through indirect reciprocity, it may be possible to elucidate how notable features of organisms (such as colour differences in bird crests) evolve, or evolution of language, which is the ultimate third party identification system. Network reciprocity may also help us understand the structure of special network topologies such as the scale-free graphs observed in many natural phenomena, as well as human social systems; in particular, how cooperation self-organizes in such networks.

---

[6] Strictly speaking PD satisfying $D_g = D_r$.

[7] This situation accords with common sense. If a game is played against the same partner each time rather than against an unknown one, both individuals should accept the cooperation option to avoid strategies leading merely to short term profit. If both individuals take the defection option P, neither will benefit long-term. Our daily behavior follows the former pattern.

[8] Many of these dynamics can be verified by simulation. Games are repeated between multiple agents in a simulated society; this approach is known as multi-agent simulation.

4.4 Dynamics Analysis of the Two-by-Two Game

This section explores how the two-by-two game dynamics differ between the four game classes explained in the previous section, i.e., Trivial with no dilemmas, PD, Chicken, and SH with dilemmas. A deductive approach, relating to the non-linear system state equations derived in Sect. 4.2, is adopted.

As before, we assume unlimited group size (i.e., infinite number of agents) existing in a well-mixed state with no social viscosities. The strategies (hands) adopted by an agent are cooperation (C) or defection (D), expressed by the following state vectors:

$$\text{Strategy C;} \quad {}^T\mathbf{e}_1 = (1 \quad 0) \tag{4.14.1}$$

$$\text{Strategy D;} \quad {}^T\mathbf{e}_2 = (0 \quad 1). \tag{4.14.2}$$

The payoff matrix of the game structure is

$$\begin{bmatrix} R & S \\ T & P \end{bmatrix} \equiv \mathbf{M}. \tag{4.15}$$

Moreover, the proportions of agents adopting strategy C and strategy D at a given time (referred to as the strategy ratio) are defined by s_1 and s_2 respectively. These strategy ratios are expressed as

$$ {}^T\mathbf{s} = (s_1 \quad s_2). \tag{4.16}$$

From the condition of simplex we get

$$s_2 = 1 - s_1. \tag{4.17}$$

The validity of Eqs. (4.14.1) to (4.17) should be understood from the following matrix equation describing the battle between two agents adopting strategy D, in which the outcome is P:

$$\pi_{DD} = (0 \quad 1) \cdot \begin{bmatrix} P & S \\ T & P \end{bmatrix} \begin{pmatrix} 0 \\ 1 \end{pmatrix} = P. \tag{4.18}$$

Equation (4.18) also computes the payoff when one strategy plays a game $\mathbf{M}$ against another with a different strategy. The expected payoff when an agent using strategy C battles with a randomly sampled agent at the present time expressed as strategy ratio $\mathbf{s}$ is

$$ {}^T\mathbf{e}_1 \cdot \mathbf{Ms}.$$

Similarly, the expected payoff when an agent using strategy D fights a randomly sampled agent at the present time expressed as strategy ratio **s** is

$$^T\mathbf{e}_2 \cdot \mathbf{Ms}.$$

The *replicator dynamics* are defined as the strategy ratio dynamics of strategy i, expressed as

$$\frac{\dot{s}_i}{s_i} = {}^T\mathbf{e}_i \cdot \mathbf{Ms} - {}^T\mathbf{s} \cdot \mathbf{Ms}. \tag{4.19}$$

The dimensionless quantity on the left hand side of (4.19), obtained by dividing $\dot{s}_i$ by the strategy ratio itself, indicates the level of change. As the reader should certainly appreciate, this quantity is determines by the extent to which the payoff for strategy i playing against the society average at a given time differs from the expected society payoff at that time. Recall how we discussed in page 80, "…even division of cooperative and defection agents (50 % cooperators & 50 % defectors), once the game is started and the strategy of the agents reviewed according to a certain set of rules after every step…" As part of this "set of rules," we investigate the evolution of the system under the replicator dynamics described in Eq. (4.19). Although other temporal dynamics can be supposed, replicator dynamics provide an adequate "set of rules" to govern evolution, for the following reason. After a game, the successful strategies (those achieving higher payoff than the average accumulated by the strategy ratio) will increase in the next time step, whereas less successful strategies will decrease. The ratio of this extent is thought to be decided by comparing with the aforementioned level of "success." In such a system, good conduct is rewarded whereas bad conduct is punished (a form of survival of the fittest). Selection mechanisms in the natural world (including human social systems) tend to operate in this manner. Alternative systems of rewarding the good and punishing the bad exist in which the response to the acquired payoffs differs from that of Eq. (4.19) may be possible. Also randomness caused by luck may enter the dynamics (i.e., poor-scoring individuals could, if lucky enough, produce offspring). In any case, we suppose replicator dynamics as the "set of rules" in the following analysis.

Substituting Eqs. (4.14)–(4.16) into Eq. (4.19) and explicitly writing the elements, we obtain

$$\{\dot{s}_1 = [(R-T) \cdot s_1 - (P-S) \cdot s_2] \cdot s_1 \cdot s_2. \dot{s}_2 = -[(R-T) \cdot s_1 - (P-S) \cdot s_2] \cdot s_1 \cdot s_2. \tag{4.20}$$

Note that when the right hand side of (4.20) = 0, the equation becomes a cubic in s_1 and s_2; that is, the system contains three equilibrium points. Two of these are self-evident:

$$\begin{pmatrix} s_1 & s_2 \end{pmatrix} = \begin{pmatrix} 1 & 0 \end{pmatrix} \equiv \mathbf{s}^*|_{\text{C-dominate}} \tag{4.21.1}$$

$$\begin{pmatrix} s_1 & s_2 \end{pmatrix} = \begin{pmatrix} 0 & 1 \end{pmatrix} \equiv \mathbf{s}^*|_{\text{D-dominate}}. \tag{4.21.2}$$

In the former, all individuals ultimately become cooperative; the latter leads to the defection state, implying C-dominant and D-dominant, respectively. The remaining equilibrium point is obtained by simultaneously solving Eq. (4.20), setting [...] on the right hand side to 0 and eliminating s_2 through Eq. (4.17) (the reader should confirm this for themselves):

$$\begin{pmatrix} s_1 & s_2 \end{pmatrix} = \begin{pmatrix} \dfrac{P-S}{P-T-S+R} & \dfrac{R-T}{P-T-S+R} \end{pmatrix} \equiv \mathbf{s}^*|_{\text{Polymorphic}}. \tag{4.21.3}$$

This third equilibrium point lies within $[0, 1]$ depending on the values of P, R, S, and T. In this case, the dynamics become polymorphic or bi-stable. Equation (4.21.3) defines an *internal equilibrium point*.

Once the three equilibrium points are obtained, the signs of the eigenvalues of the Jacobian matrix at each equilibrium point are determined, and the equilibrium points are assessed as sink, source, or saddle.

To this end, we re-write Eq. (4.20) as follows:

$$\dot{s}_1 \equiv f_1(s_1, s_2) \tag{4.22.1}$$

$$\dot{s}_2 \equiv f_2(s_1, s_2). \tag{4.22.2}$$

From Eq. (4.17), we observe that $f_1 = -f_2$. Hence, the Jacobian (4.12) is calculated as

$$\left\{ \begin{aligned} & \frac{\partial f_1}{\partial s_1} = -\frac{\partial f_2}{\partial s_1} = 3(-R+S+T-P)s_1{}^2 \\ & \qquad +2(R-2S-T+2P)s_1 + S - P. \\[2mm] & \frac{\partial f_1}{\partial s_2} = -\frac{\partial f_2}{\partial s_2} = -3(-R+S+T-P)s_1{}^2 \\ & \qquad -2(R-2S-T+2P)s_1 - S + P. \end{aligned} \right. \qquad \begin{aligned} (4.23.1) \\[8mm] (4.23.2) \end{aligned}$$

The reader is encouraged to verify these equations. The Jacobian matrix

$$\mathbf{J} = \begin{bmatrix} \dfrac{\partial f_1}{\partial s_1} & \dfrac{\partial f_1}{\partial s_2} \\[2ex] \dfrac{\partial f_2}{\partial s_1} & \dfrac{\partial f_2}{\partial s_2} \end{bmatrix} = \begin{bmatrix} \dfrac{\partial f_1}{\partial s_1} & \dfrac{\partial f_1}{\partial s_2} \\[2ex] -\dfrac{\partial f_1}{\partial s_1} & -\dfrac{\partial f_1}{\partial s_2} \end{bmatrix} \quad \text{is a } 2 \times 2 \text{ matrix, so its eigenvalues}$$

$\left(0 \text{ and } \frac{\partial f_1}{\partial s_1} - \frac{\partial f_1}{\partial s_2}\right)$ are easily obtained using senior school mathematics (readers should try to recall and apply the eigenvalue calculations from their maths textbooks). Since 0 is unsigned, we need only obtain the sign of $\frac{\partial f_1}{\partial s_1} - \frac{\partial f_1}{\partial s_2}$ to establish the equilibrium conditions. Explicitly, these eigenvalues are

$$\lambda = \frac{\partial f_1}{\partial s_1} - \frac{\partial f_1}{\partial s_2} = 6(-R + S + T - P)s_1{}^2 \tag{4.24}$$
$$+ 4(R - 2S - T + 2P)s_1 + 2(S - P).$$

1. The necessary and sufficient condition for the equilibrium point $\mathbf{s}^*|_{\text{C-dominate}}$ to be sink is $\lambda < 0$ when substituting $(s_1 \quad s_2) = (1 \quad 0)$ into Eq. (4.24). The following conditions are sought:

$$T - R = D_g < 0. \tag{4.25}$$

2. The necessary and sufficient condition for the equilibrium point $\mathbf{s}^*|_{\text{D-dominate}}$ to be a sink is $\lambda < 0$ when substituting $(s_1 \quad s_2) = (0 \quad 1)$ into Eq. (4.24). We now require that

$$P - S = D_r > 0 \tag{4.26}$$

3. The necessary and sufficient conditions for the equilibrium point $\mathbf{s}^*|_{\text{Polymorphic}}$ to be a sink is $\lambda < 0$ with $(s_1 \quad s_2) = \left(\dfrac{P - S}{P - T - S + R} \quad \dfrac{R - T}{P - T - S + R} \right)$ substituted into Eq. (4.24). Noting that $\lambda = 2\dfrac{(R - T)(P - S)}{R - S - T + P}$, we seek the following conditions:

$$P < S \wedge R < T \Leftrightarrow P - S = D_r < 0 \wedge T - R = D_g > 0. \tag{4.27}$$

Table 4.1 2 × 2 game dynamics derived analytically

Game class	Phase	Nash equilibrium	Sign of D_g	Sign of D_r	Each point sink, source, or saddle		
					(1,0)	(0,1)	$\left(\dfrac{D_r}{D_g-D_r} \quad \dfrac{-D_g}{D_r-D_g}\right)$
PD	**D**-dominate	(0,1)	+	+	Source	Sink	Saddle
Chicken	Polymorphic	$\left(\dfrac{D_r}{D_g-D_r} \quad \dfrac{-D_g}{D_r-D_g}\right)$	+	−	Source	Source	Sink
SH	Bi-stable	(0,1) or (1,0)	−	+	Sink	Sink	Source
Trivial	C-dominate	(1,0)	−	−	Sink	Source	Saddle

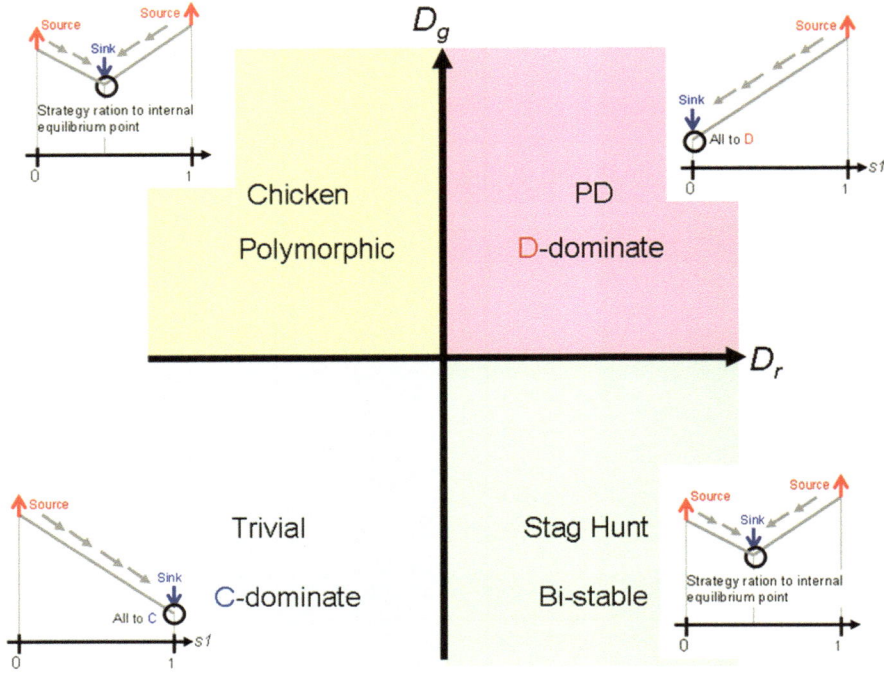

Fig. 4.8 Phase diagram of dynamics classified by D_g and D_r of two-by-two game and a summary of dynamics of each game class

The above conditions are summarized in Table 4.1, with the following substitution:

$$\mathbf{s}^*|_{\text{Polymorphic}} = \left(\frac{P-S}{P-T-S+R} \quad \frac{R-T}{P-T-S+R} \right) = \left(\frac{D_r}{D_g - D_r} \quad \frac{-D_g}{D_r - D_g} \right).$$

Defining D_g and D_r in Eq. (4.13), the four game classes were established as PD, Chicken, SH, and Trivial (see Fig. 4.4). Here, these divisions are represented by the difference between the signs of the three equilibrium points.

In PD, $\mathbf{s}^*|_{\text{C-dominate}}$ and $\mathbf{s}^*|_{\text{D-dominate}}$ are source and sink, respectively; hence, regardless of the initial cooperation proportion in [0, 1] the ultimate state is one of complete defection at $t \rightarrow \infty$.

In Chicken, $\mathbf{s}^*|_{\text{C-dominate}}$ and $\mathbf{s}^*|_{\text{D-dominate}}$ are both sources. In this case $\mathbf{s}^*|_{\text{Polymorphic}}$ (value in [0, 1]) is a sink, so regardless of initial cooperation proportion, as $t \rightarrow \infty$, the system settles to the internal equilibrium point $\mathbf{s}^*|_{\text{Polymorphic}}$. As previously mentioned, this state does not imply that specific agents are fixed into cooperation or defection strategies, but that when the infinitely large group is viewed as a whole, the proportions of cooperation and defection players are (dynamically) steady.

In SH, the internal equilibrium point $s^*|_{Polymorphic}$ is a source, while $s^*|_{C\text{-dominate}}$ and $s^*|_{D\text{-dominate}}$ are both sinks. Therefore if the initial proportion of cooperative players is smaller (or larger) than $s^*|_{Polymorphic}$, the ultimate state is pure defection, (or pure cooperation), and the system is bi-stable.

In Trivial, $s^*|_{C\text{-dominate}}$ is a sink and $s^*|_{D\text{-dominate}}$ is a source, so regardless of the initial cooperation proportion, the pure cooperation state is inevitable. For this reason, Trivial is a game with no dilemmas.

The above discussion is summarized schematically in Fig. 4.8.

Here, we have fully characterized the 2×2 replicator dynamics, expressed as non-linear cubic equations.

The following is provided for interest only. A two by two game has two strategies, so the dynamics are relatively simple, and one of the equilibrium points inevitably acts as a sink. If the number of strategies is increased, more degrees of freedom are introduced, leading to *perturbation* dynamics (which display periodic behavior) or *chaos* (which is deterministic but unpredictable). The interested reader should take a look at related literatures [4, 5].

References

1. Hardin G (1968) Tragedy of the commons. Science 162(3859):1243–1248
2. Nowak MA (2006) Five rules for the evolution of cooperation. Science 314:1560–1563
3. Hamilton WD (1963) The evolution of altruistic behavior. Am Nat 97:354–356
4. Weibull JW (1997) Evolutionary game theory. MIT, Cambridge
5. Nowak MA (2006) Evolutionary dynamics. Belknap Press of Harvard University Press, Cambridge

Biography of the Author

Professor Jun Tanimoto, Dr. Eng., was born in 1965 in Fukuoka, but had grown up in Yokohama; graduated from Department of Architecture, Undergraduate School of Science & Engineering, Waseda University in 1988, also completed his master's program in 1990, finally got his doctoral degree from Waseda Univ. in 1993. He started his professional carrier as Research Associate at Tokyo Metropolitan University in 1990, moved to Kyushu University with promotion to Assistant Professor (Senior Lecturer) in 1995, became Associate Professor in 1998, and has been responsible as Professor from 2003 (taking charge of Head of Laboratory of Urban Architectural Environmental Engineering); also served as Visiting Professor at National Renewable Energy Laboratory (NREL), USA; University of New South Wales, Australia; and Eindhoven University of Technology, Netherlands. He was awarded with Award of Society of Heating, Air-Conditioning and Sanitary Engineers of Japan (SHASE); Fosterage Award of Architectural Institute of Japan (AIJ); Award of AIJ; and IEEE CEC2009 Best Paper Award. His world-wide dedication covers diverse activities; Editor at several international journals, Committee member in many conferences, public services such as Expert at IEA Solar Heating and Cooling Program Task 23, and so on. Surprisingly, he has been active as a painter as well as a novelist, awarded with many fine art prizes and literature ones. You can visit http://ktlabo.cm.kyushu-u.ac.jp/ for more information.

Index

A
Amplification coefficient, 27–31
Annual steady calculation, 18
Atmospheric permeability, 100

B
Backward FDM, 10
Berlage formula, 100, 101
Bi-stable, 115, 119, 121, 123
Bouguer's equation, 100

C
C-dominate, 115, 119–123
Chicken (Snow Drift, Hawk–Dove) game,
  110, 114
Coexistence, 115
Composite conductance, 13, 61, 62
Control volume method (CDM), 9–16,
  19, 71
Courant number, 28
Crank–Nicolson method, 10, 17, 19, 20, 23–26,
  28, 30, 48, 102

D
Daily steady-state solution, 58, 69, 70
D-dominate, 114, 119–123
Declination of the sun, 99
Diffusion coefficient, 6
Diffusion equation, 8, 15, 43, 44
Diffusivity, 6
Direct diffuse separation (for solar
  radiation), 97
Discretization, 4, 9–24, 26–30, 54, 60, 61, 63,
  70–72, 78, 87, 107, 108

E
Eigenvalue, 21, 23–27, 106–111, 119, 120
Elliptic equation, 9
Environmental system, 1–4, 97
Equation of time, 98, 99
Expansion capacitance matrix, 45
Expansion conductance matrix, 46
Explicit method, 19, 21, 72, 75, 78

F
Fair Pareto optimization, 112, 114
Finite difference method (FDM), 9, 10, 15,
  16, 19, 40
Finite element method (FEM), 4, 9, 15, 69–78
Forward FDM, 10
Fourier's law, 7, 8

G
Gauss divergence theorem, 74, 79

H
Heat
  capacitance matrix, 15, 16, 31–33, 35, 51,
    53, 61, 70
  capacity matrix, 15
  conductance matrix, 16, 32, 33, 35, 37, 40,
    52, 53, 61, 87
  moisture transfer equation, 41–48
Horizontal surface diffuse irradiance, 98–101
Hour angle, 98
Human-environmental-social system, 1, 2
Hygroscopic, 41, 42, 45
Hyperbolic equation, 9, 10
Hyperbolic relationships, 28